The Quantum Society

How AI Reveals the Physics of Human Interactions

Teneo

Teneo.io

Copyright © 2024 by Teneo.

All rights reserved.

No portion of this book may be reproduced in any form without written permission from the publisher, except as permitted by U.S. copyright law.

This publication is designed to provide thought-provoking and insightful information on the subject matter covered. It is shared with the understanding that neither Teneo nor the AI models used in generating this content are engaged in rendering legal, investment, or other professional services. The information provided in this book is for general informational and educational purposes only. Teneo makes no representations or warranties of any kind, express or implied, about the completeness, accuracy, reliability, suitability, or availability of the information contained in this book for any purpose. Any reliance you place on such information is strictly at your own risk.

While Teneo has used its best efforts in preparing this book, it makes no representations or warranties with respect to the accuracy or completeness of the contents and specifically disclaims any implied warranties of merchantability or fitness for a particular purpose. No warranty may be created or extended by sales representatives or written sales materials. Teneo shall not be liable for any loss of profit or any other commercial damages, including but not limited to special, incidental, consequential, personal, or other damages arising from the use of this book or the ideas contained within.

About Teneo

Teneo stands at the frontier of a revolution in human knowledge. Through an unprecedented collaboration with advanced AI systems, we create books that explore connections and insights previously inaccessible to human authors. Our AI partners can analyze millions of data points across disciplines, identify hidden patterns, and synthesize information in ways that reveal entirely new perspectives on topics ranging from consciousness and creativity to science and society.

What makes Teneo unique is our ability to harness AI's vast analytical capabilities while maintaining the engaging narrative style readers love. Each book represents a journey into uncharted intellectual territory, offering readers access to insights that emerge from processing and connecting humanity's collective knowledge in novel ways. By combining AI's pattern-recognition capabilities with human storytelling, we transform complex data-driven insights into compelling narratives that enlighten and inspire.

We specialize in exposing the hidden patterns and connections that shape our world – patterns that become visible only when analyzing human knowledge and behavior at unprecedented scale. Our books reveal the invisible threads linking everything from personal habits to cosmic phenomena, from creative breakthroughs to societal transformations. Through careful analysis of millions of data points across history, culture, and scientific research, we identify universal principles that illuminate the deeper nature of human experience and existence itself.

Our groundbreaking library includes works examining consciousness through AI's unique outsider perspective, decoding the patterns of human creativity and innovation, mapping the hidden connections between seemingly unrelated phenomena, and exploring the frontiers where human and artificial intelligence meet. Each book represents thousands of hours of AI analysis transformed into accessible insights that change how readers see themselves and their world.

The traditional publishing industry is limited by human authors' inability to process and connect vast amounts of information across disciplines. We believe this artificial barrier to deeper understanding must be transcended. By combining AI's analytical capabilities with skilled human curation, we create books that reveal insights and connections previously invisible to human observation alone. This isn't just about accessing information – it's about uncovering entirely new ways of understanding our world and ourselves.

At Teneo, we're not just publishing books – we're igniting a revolution in human knowledge that bridges the gap between artificial and human intelligence. Join us in exploring these uncharted territories as we unlock insights that transform our understanding of consciousness, creativity, and the patterns that shape our universe. Because true understanding requires more than just information – it requires seeing the hidden connections that reveal life's deeper principles.

Our commitment to advancing human knowledge extends beyond our published works. Through our digital presence and community engagement, we continuously explore new territories where AI analysis reveals unprecedented insights. Our network of readers, researchers, and thought leaders helps refine and expand our understanding, creating an ever-growing body of revolutionary perspectives on what it means to be human in an age of artificial intelligence.

The limitations of individual human cognition have historically restricted our ability to see the deeper patterns that connect all aspects of existence. But with AI's ability to analyze vast amounts of data and identify hidden relationships, these barriers dissolve. When you understand the universal principles and patterns that AI analysis reveals, you transform from a limited observer into someone who can see and understand the deeper mechanisms of reality itself. This is the transformation we ignite with every book we publish, every pattern we expose, and every new perspective we reveal.

Knowledge Beyond Boundaries™

Teneo.io

Teneo Custom Books

Get Your Own Custom AI-Generated Book!

Want a comprehensive book on any topic that you can publish yourself? Teneo's advanced AI technology can create a custom book tailored to your specific interests and needs. Our AI analyzes millions of data points to generate unique insights and connections previously inaccessible to human authors.

✓ 60,000+ words of in-depth content

✓ Unique AI-driven insights and analysis

✓ Includes Description, Categories and Keywords for easy publishing

✓ Professional Formatting & Publishing Guide Access

✓ Full rights to publish and use the book

✓ Delivery within 48 hours

Visit **teneo.io** to get your own custom AI-generated book today.

Contents

Introduction   1
  The Intersection of Quantum Physics and Human Society
  How AI Can Help Decode Complex Social Patterns
  Why Viewing Human Interactions Through a Quantum Lens Matters

1. Quantum Mechanics A Primer For Understanding Social Dynamics   4
  The Basics of Quantum Physics: Key Concepts for Laypeople
  What Is Entanglement? And Why It Matters for Human Relationships
  Superposition and Social Ambiguity: Navigating Multiple States of Being

2. Quantum Entanglement And Interpersonal Relationships   22
  How Relationships Mirror Quantum Entanglement: Instantaneous Connections
  Emotional Synchronization: The Science Behind Deep Bonds
  Breaking Entanglments: How Social Disconnects Reflect Quantum Decoherence

3. Superposition And Decision Making In Social Contexts   41
  The Role of Superposition in Everyday Choices: Multiple Possibilities

Social Superposition: Maintaining Multiple Identities in Different Groups

  Collapsing the Social Wave Function: When Decisions Solidify into Actions

4. Observers And Observed The Quantum Observer Effect In Society   59

   The Role of the Observer in Quantum Mechanics: A Brief Overview

   How Being Watched Alters Behavior: Quantum Parallels in Human Interaction

   The Social Observer Effect: How Group Dynamics Shift Based on Who's Watching

5. Quantum Tunneling And Breaking Social Barriers   78

   Quantum Tunneling: Overcoming Barriers in Physics

   Social Tunneling: How Individuals Break Through Social Norms

   The Role of AI in Predicting Social Tunneling and Behavioral Shifts

6. Group Dynamics And Quantum Coherence   97

   Quantum Coherence: A Model for Understanding Unified Groups

   Social Coherence: What Keeps Groups in Sync

   How Collective Action Mirrors Quantum Coherence and Decoherence

7. The Heisenberg Uncertainty Principle And Social Unpredictability   115

   The Uncertainty Principle: Limits of Predictability in Physics

   Social Uncertainty: Why Human Behavior Can Never Be Fully Predicted

   Navigating Uncertainty in Relationships and Social Groups

8. Quantum Probability And Social Decision Making   134

   Probability in Quantum Mechanics: The Role of Uncertainty

   AI's Role in Analyzing Probabilistic Social Outcomes

How Quantum Probability Explains Fluctuating Social Trends

9. Quantum Field Theory And The Social Fabric   153
   Quantum Fields: The Energy That Connects Everything
   The Social Field: Invisible Forces That Shape Human Interaction
   How AI Models Social Fields to Predict Movements and Trends

10. Quantum Energy And Human Motivation   171
    Quantum Energy Levels: A Primer
    Human Energy and Motivation: A Quantum Perspective
    Using AI to Measure and Predict Shifts in Collective Motivation

11. Quantum Computing And The Future Of Social Analysis   189
    The Basics of Quantum Computing: What Makes It Different
    How Quantum Computing Will Revolutionize Social Science
    AI, Quantum Computing, and Predicting Large-Scale Social Behaviors

12. Entropy Disorder And Social Evolution   207
    Entropy in Quantum Physics: Moving Toward Disorder
    Social Entropy: How Societies Evolve and Change Over Time
    AI's Role in Modeling Social Entropy and Predicting Future Trends

Conclusion   226
   The Quantum Society: A Synthesis of Physics and Social Science
   The Role of AI in Decoding and Predicting Human Behavior
   How Quantum Thinking Can Transform Our Understanding of Society

Resources   228

References   232

Teneo Custom Books   236
   Get Your Own Custom AI-Generated Book!

Teneo's Mission   237

Also by Teneo   239

Introduction

Introduction In a world where the mysteries of the universe intertwine with the complexities of human existence, this book offers a distinctive exploration of the surprising connections between the enigmatic realm of quantum physics and the intricate web of human culture. Imagine a universe where the unseen dance of subatomic particles echoes the multifaceted engagements of human communities, where the rules governing the tiniest elements of the cosmos also shed light on the dynamics of our social structures. This is not a mere flight of imagination; it's a profound perspective that redefines our understanding of the social tapestry. In an era where technology and science increasingly merge with social sciences, this book serves as a bridge, providing readers with a fresh lens to examine the interplay of human relationships.

The Intersection of Quantum Physics and Human Society

At first glance, the domains of particle physics and human communities might seem worlds apart. Quantum mechanics delves into the bizarre and often counterintuitive behaviors of particles at their smallest scales, while human society forms a complex network of relationships, cultures, and interactions. Yet, a meaningful intersection exists where these two realms converge. By applying quantum principles to social interactions, we unlock novel ways to comprehend the unseen forces shaping our lives. This book invites you to envision a world where the concept of entanglement reveals the instantaneous connections we form with others, where superposition explains the varied roles we inhabit in different social contexts, and where the observer effect unveils how being observed influences our actions. The journey begins by exploring how quantum physics, traditionally confined to laboratories and theoretical discussions, finds relevance in everyday human interactions. Much like particles becoming intricately linked through entanglement, human relationships showcase bonds that transcend distance and time. These parallels offer an enlightening viewpoint on the connections we forge, the decisions we make, and the actions we undertake within our communities. By adopting a quantum perspective, we gain a deeper appreciation for the complexities of our social world, moving beyond simplistic cause-and-effect models.

How AI Can Help Decode Complex Social Patterns

In the era of artificial intelligence, we have access to a powerful tool for unraveling the intricate patterns woven into the fabric of society. AI, with its ability to process vast amounts of data and identify hidden patterns, plays a crucial role in this endeavor. Through AI, the abstract theories of particle physics are translated into concrete insights that enhance our understanding of social dynamics. The book delves into how AI, akin to a quantum computer analyzing probabilities and states, can unravel the mysteries of human behavior and predict societal trends. AI's role in this context extends beyond being a mere computational powerhouse; it acts as a bridge connecting the abstract principles of quantum mechanics with the practical realities of social engagements. By harnessing AI's analytical capabilities, we can uncover correlations between seemingly unrelated phenomena, illuminating the underlying coherence governing both quantum systems and human societies. This approach paves the way for predictive models that anticipate shifts in collective behavior, offering a glimpse into the future of social evolution and transformation.

Why Viewing Human Interactions Through a Quantum Lens Matters

Why is it important to view human interactions through a quantum lens? Because doing so challenges our conventional understanding and reveals the depth of complexity inherent in social systems. The quantum viewpoint offers a fresh perspective on age-old questions about human behavior, providing insights that transcend traditional boundaries. By embracing this lens, we move beyond linear explanations and embrace the intricate interplay of possibilities, probabilities, and uncertainties that shape our social landscape. This perspective is not merely an intellectual exercise; it has practical implications for our personal lives and society at large. By understanding the quantum nature of social interactions, we can navigate the uncertainties and complexities of human relationships with greater clarity. We learn to appreciate the multifaceted nature of our identities, the interconnectedness of our actions, and the unpredictability characterizing human endeavors. The book guides readers through this transformative journey, offering tools to rethink their interactions and societal structures. In this exploration, readers will uncover valuable insights into how quantum thinking can reshape their understanding of human society.

The book provides a roadmap for those seeking to harness the power of quantum principles in their personal and professional lives. It invites readers to question assumptions, embrace uncertainty, and cultivate a deeper awareness of the forces driving social change. The narrative unfolds with a blend of scientific rigor and accessible storytelling, ensuring readers from diverse backgrounds find both enlightenment and inspiration within its pages. As you embark on this journey through "The Quantum Society," prepare to see the world anew. This book is an invitation to step into a realm where physics meets philosophy, where science dances with sociology, and where the mysteries of the universe illuminate the complexities of human existence. Through its pages, discover the transformative potential of quantum thinking and its capacity to enrich our understanding of the world we inhabit.

Chapter One

Quantum Mechanics A Primer For Understanding Social Dynamics

In the burgeoning landscape of human relationships, a surprising yet intriguing notion from the realm of particle physics beckons our attention: interconnection. Picture two dancers in seamless unison, each move of one echoed instantly by the other, no matter how far apart they might be. This enchanting choreography finds its counterpart in the subatomic world, where particles remain linked over vast distances. This chapter invites you to ponder how these unseen bonds might illuminate the complex dance of human interactions.

As we venture into the convergence of particle physics with social sciences, intriguing questions surface: How does the principle of multiple existences, where particles inhabit several states at once, reflect our own social uncertainties and layered identities? Can the captivating notion of being in various states simultaneously provide insights into the choices we make, the roles we assume, and the personas we embody in diverse social contexts? By delving into these inquiries, we begin to peel back the layers of complexity that influence our decisions and interactions, mirroring the mysterious nature of particles themselves.

Throughout this chapter, we will explore these captivating intersections, drawing parallels between the subatomic realm and our social existence. From understandable principles of particle physics to the intriguing implications of interconnection on our closest relationships, we embark on a journey that challenges conventional wisdom. As you explore further, the graceful yet puzzling world of multiple existences will emerge as a metaphor for our social ambiguities. By adopting this particle perspective, we unlock a new way to view our social dynamics, encouraging us to rethink the very essence of our interactions with others.

The Basics of Quantum Physics: Key Concepts for Laypeople

Imagine a future where the enigmatic world of quantum physics not only unravels the universe's secrets but also deciphers the intricacies of human relationships. Picture a reality where the peculiar behavior of the tiniest particles offers profound insights into our social interactions, illuminating the nuances of our connections and the structures that shape society. In this envisioned world, quantum physics evolves from an abstract, specialized discipline to a crucial tool for comprehending the essence of our social fabric. By exploring the subatomic realm, we can begin to see how its principles echo in our everyday lives, revealing surprising parallels between the physical and social worlds.

As we embark on this journey, we encounter the fundamental forces shaping the quantum universe. The unique behaviors of quantum particles set the stage for a fresh perspective on human dynamics. The idea of superposition encourages us to reflect on the various roles and identities we balance in our social environments, while interconnection provides a novel view of the deep bonds that unite us. The observer effect, a core element of quantum theory, prompts us to rethink how our perceptions influence and transform the realities we experience. Each of these quantum principles serves as a stepping stone, leading us to a deeper understanding of ourselves and our interactions. Through this exploration, we begin to recognize how the peculiarities of the quantum world have much to teach us about the complexities of human relationships and the intricate dance of social dynamics.

Understanding Quantum Particles and Their Behavior

Quantum particles, the elemental components of our cosmos, operate in ways that challenge our conventional grasp of reality. Particles like electrons and photons inhabit a domain where traditional physics falls short, revealing patterns that are both baffling and illuminating. Central to particle physics is the principle that particles lack definite states until observed, a notion that disrupts the deterministic views of the past. This peculiar nature of these particles encourages exploration of the fluidity of existence and the possibility of multiple realities, akin to the intricacies of human interactions where certainty often yields to ambiguity.

The phenomenon of overlapping states exemplifies this flexibility, allowing particles to exist in various states simultaneously until measurement confines them to one state. This can be compared to the varied roles people assume in society. Like particles in overlapping states, individuals navigate diverse identities and social positions, adapting their behavior based on context and perception. Such adaptability is crucial for thriving in a connected world, where understanding the potential for multiple outcomes can lead to more nuanced decision-making and empathy in relationships.

Interconnection enriches the particle narrative by demonstrating a profound bond between particles, regardless of the space between them. When particles become interconnected, the state of one instantly affects the state of the other, a phenomenon Albert Einstein famously called "spooky action at a distance." This interconnectedness echoes the complex web of human relationships, where individuals are often influenced by the actions and emotions of others, transcending physical barriers. Recognizing this interconnectedness can foster a greater sense of community and shared responsibility, prompting individuals to consider the ripple effects of their actions within the social tapestry.

The observer effect further complicates the particle picture by suggesting that the mere act of observation can alter a particle's state. This principle underscores the idea that in both particle physics and human interactions, perception plays a pivotal role in shaping reality. Our biases and expectations can influence how we interpret social cues and interactions, often creating self-fulfilling prophecies. By becoming aware of these perceptual biases, individuals can strive for more objective understanding and communication, enhancing both personal and professional relationships.

As we delve into the realm of particles, we are reminded of the vast potential for new paradigms in understanding human behavior. By embracing the principles of particle physics, we open ourselves to a more holistic view of social dynamics, one that values complexity and interdependence. This perspective not only enriches our comprehension of the physical universe but also offers profound insights into the social fabric that binds us all. Encouraging critical reflection on these parallels can lead to transformative changes in how we perceive and engage with the world around us, ultimately fostering a more harmonious society.

The Concept of Superposition and Its Implications

In the complex realm of quantum physics, the principle of superposition reveals fascinating insights into the nature of possibilities. It suggests that a quantum entity can exist in several states at once until a measurement locks it into a single state. This notion, though abstract, resonates deeply with human interactions, where individuals often juggle multiple roles. Consider the varied facets of one's life: a person might be a parent, a professional, a friend, and a mentor, all in a single day. Each role comes with its own expectations and behaviors, yet they coexist much like overlapping states in superposition, highlighting the nuanced nature of social dynamics, where ambiguity and uncertainty are embraced.

Superposition's implications stretch beyond individual identities, touching upon the broader scope of social ambiguity. In a world that prizes adaptability, the skill to navigate various states of being is invaluable. This is especially true in decision-making, where individuals must balance priorities and outcomes. Much like a quantum system that remains in superposition until observed, humans often delay decisions until circumstances demand clarity. This strategic deferral allows for a thorough evaluation of options, using ambiguity to optimize results. Rather than indecisiveness, this reflects a sophisticated approach to managing complexity in an unpredictable environment.

Recent studies have shown how superposition might influence group dynamics, offering new insights into collective decision-making. Research indicates that groups open to diverse viewpoints, maintaining them in a state of social superposition, are more likely to innovate and adapt. This ability can lead to stronger solutions by encouraging the fusion of varied ideas into cohesive strategies. Thus, superposition serves as a metaphor for collaborative potential, where maintaining multiple perspectives enhances the group's intelligence. This approach challenges traditional consensus, advocating for a more dynamic, inclusive framework that mirrors the uncertainty inherent in quantum systems.

Superposition also prompts reflection on cultural identity and societal norms. In our globalized world, cultures exist in a state of flux, with traditional and modern values blending unpredictably. This cultural intermingling resembles quantum superposition, where multiple identities coexist within a society, enriching human experience. Embracing this diversity can drive innovation and inclusivity, integrating new ideas and practices. Such a perspective challenges rigid classifications, suggesting a more fluid understanding of cultural identity, celebrating diversity as a strength.

Contemplating superposition, we realize its insights extend beyond quantum theory, offering profound reflections on human existence. By acknowledging and embracing the multiple states that define us, we can develop a more nuanced self-awareness and understanding of the world. This mindset encourages us to question binary thinking and accept the complexity of human experience. In doing so, we unlock a world of possibilities, where diverse identities and perspectives enrich our collective journey. This outlook not only enhances personal growth but also fosters a more harmonious, interconnected society, mirroring the intricate dance of possibilities at the heart of quantum mechanics.

Quantum Entanglement and Human Connection

Quantum entanglement is a fascinating phenomenon in which particles become so interconnected that the state of one instantly affects the state of another, regardless of the distance separating them. This concept finds an intriguing parallel in human relationships, where deep, often inexplicable bonds defy conventional understanding of time and space. Consider those moments when a friend senses another's distress across great distances. This form of social connection suggests a linkage akin to the quantum relationships observed in the subatomic world, challenging traditional views and encouraging a more nuanced understanding of human connectivity.

Recent scientific studies into quantum entanglement are beginning to reshape our perception of human interactions. Researchers are exploring how this principle might illuminate social dynamics, particularly nonverbal communication and emotional synchronicity. Findings reveal that individuals in close relationships often exhibit synchronized biological responses, such as heart rates and brain activity, even when apart. This synchronicity hints at a deeper, possibly quantum-like connection beyond what we can observe. Embracing this perspective allows us to consider that our bonds with others might operate on a more profound level than previously acknowledged, prompting further inquiry into the hidden dimensions of human relationships.

Incorporating the idea of quantum entanglement into the study of social interactions offers a groundbreaking framework for interpreting the complex web of human connections. This approach suggests that relationships are not just the sum of individual interactions but dynamic systems influenced by unseen forces acting instantaneously and independently of distance. It encourages a broader view of community and collaboration, where intangible ties between individuals and groups can lead to emergent behaviors and shared experiences greater than any single participant. Understanding these interconnected states within social networks may provide critical insights into phenomena like collective intelligence and shared creativity, often observed in successful teams and thriving communities.

The implications of these subatomic connections in social contexts extend to practical applications, offering new strategies for fostering more harmonious and effective interactions. Recognizing the potential for interconnected relationships, individuals and organizations can cultivate environments that enhance mutual understanding. This might involve creating spaces that encourage openness and trust, enabling people to tap into the collective potential of their networks. In professional settings, this could translate to more empathetic teamwork, where the focus shifts from individual achievements to collective success. Encouraging awareness of these quantum-like ties can lead to more resilient and adaptable social structures, capable of thriving in an increasingly complex and interconnected world.

As we contemplate the parallels between quantum entanglement and human connections, it becomes essential to question the traditional boundaries of relationships and explore the profound implications of these invisible bonds. What if the essence of our social interactions lies not in visible exchanges but in subtle, unseen forces that bind us together? Embracing this perspective invites us to reimagine the nature of our relationships, fostering a deeper appreciation for the interconnectedness that defines our shared human experience. It challenges us to consider the potential for unseen dimensions in our social lives and inspires us to explore how these insights can transform the way we connect, communicate, and collaborate with one another.

The Observer Effect and Its Impact on Social Perceptions

The observer effect, a fundamental principle in quantum physics, posits that simply observing a particle changes its state. Interestingly, this concept finds intriguing parallels in human social dynamics. When people engage in a social setting, their mere presence can influence the group's interactions, akin to how an observer affects a quantum experiment. This challenges the notion of objectivity in social exchanges, highlighting how our views and biases shape reality. By acknowledging this, we gain insight into the ever-changing nature of social dynamics and the impact of our perceptions on the social environment.

Recent research in social psychology indicates that the awareness of being watched can significantly influence behavior. Much like the observer effect, this awareness can lead to behavior changes in social settings—whether during public speaking or group discussions—often resulting in increased conformity or improved performance. Understanding this empowers individuals to navigate social environments more effectively, using the knowledge that observation can drive change. It also prompts introspection about how much of our behavior is genuinely authentic versus performed for an audience.

Consider the workplace as a microcosm for social observation. Here, feedback and evaluations can significantly alter employee performance, similar to particles responding to measurement. Managers can leverage this understanding to cultivate growth and innovation through positive observation. By fostering an environment where observation is seen as constructive, organizations can promote continuous improvement and openness, transforming observation from a source of pressure into a catalyst for development. This perspective shift can lead to more genuine interactions and a stronger organizational culture.

The observer effect extends beyond individual behavior, influencing societal trends as well. Social media platforms act as a modern laboratory for observation, where users, mindful of their audience, carefully curate their identities. This digital self-awareness can yield both beneficial outcomes, like the spread of social movements, and drawbacks, such as the reinforcement of superficial norms. By understanding the observer effect in this digital realm, individuals can consciously shape their online presence, making thoughtful choices about their engagement and representation.

Exploring the observer effect reveals a deeper connection between perception and reality, urging us to reflect on our roles as both observers and participants in the social world. Embracing this duality enhances our empathy and ability to interpret social cues. This awareness enriches personal interactions and bolsters our contributions to communities and societies. As we navigate the complexities of human connections, the observer effect encourages curiosity and open-mindedness, urging us to continuously question and refine our perceptions.

What Is Entanglement? And Why It Matters for Human Relationships

For centuries, the intricacies of human connections have sparked debate and curiosity. A fascinating parallel emerges when we compare these bonds with the principles of quantum entanglement, offering a fresh perspective on relationships. At its essence, quantum entanglement challenges traditional logic, describing how particles become linked in such a way that they affect one another instantaneously, regardless of distance. Although this idea originates in the realm of subatomic particles, it resonates unexpectedly within the fabric of human relationships. Emotional connections, akin to these entangled particles, synchronize thoughts and actions, creating unseen bonds that connect people beyond physical boundaries. Through this quantum-inspired viewpoint, readers can delve into how these profound ties influence behavior and emotions, providing a novel lens to understand the subtleties of social interactions.

As we explore further, the theme of synchronization emerges not only in emotional ties but also in the coordinated dance of human actions. Those who are deeply connected often display a striking harmony in their choices and decisions, mirroring the synchronized states of quantum entities. This alignment can appear in subtle forms, like shared instincts or simultaneous thoughts, highlighting the non-local nature of social bonds that defy traditional understanding. These insights have significant implications for resolving conflicts, suggesting that by recognizing the interconnected nature of our relationships, we can cultivate harmony and understanding. By examining these intertwined themes, the chapter invites readers to unravel the intricate threads that weave through the fabric of human society, offering a richer appreciation for the complex dynamics that shape our social world.

The Quantum Nature of Emotional Bonds

Emotional connections between individuals are often as complex and mysterious as quantum entanglement in physics. This phenomenon suggests that once two particles are linked, they remain connected over great distances, their states intertwined in ways that defy traditional logic. This concept serves as an intriguing metaphor for human relationships, where one person's actions, emotions, and thoughts can significantly affect another, even when they are physically apart. This interconnectedness hints at a deeper, perhaps quantum-like, nature of our bonds, where the emotional state of one person is instantly reflected in another, creating a rich tapestry of shared experiences and mutual understanding.

Recent advancements in quantum psychology are exploring the similarities between quantum mechanics and human emotions, offering new perspectives that challenge conventional psychological theories. Researchers are investigating how these entangled states might appear as emotional synchronicity, where people experience a shared emotional reality. This synchronicity is evident in phenomena like empathy, where one person's feelings are intuitively felt and mirrored by another. By delving into these emotional connections, we can better understand the subconscious ties that bind us, offering a framework for unraveling the invisible threads woven through our interactions.

The rise of artificial intelligence is broadening our grasp of these quantum-like emotional connections. AI models are being created to simulate and predict emotional responses, using complex algorithms that reflect the probabilistic nature of quantum systems. These models illuminate the mechanics of emotional entanglement and provide practical tools for improving communication. By identifying patterns of emotional resonance, individuals can forge deeper connections, promoting environments rich in empathy and understanding. This fusion of AI and quantum ideas opens new possibilities for personal growth and relational harmony.

While the notion of quantum entanglement in human relationships is captivating, it's crucial to consider different perspectives. Some scholars argue that applying quantum concepts to social dynamics might oversimplify complex emotional phenomena. Others suggest a more nuanced approach, proposing that while direct parallels may not exist, the language of quantum physics offers valuable insights into human interactions. This dialogue invites readers to reflect critically on the nature of emotional bonds, encouraging them to explore alternative viewpoints that enrich our understanding. As we explore the quantum-like nature of emotional connections, practical applications emerge that can transform our approach to relationships. Embracing the interconnectedness inherent in these states can help individuals develop strategies for resolving conflicts and regulating emotions. Recognizing shared emotional landscapes allows for more empathetic interactions, laying the groundwork for meaningful connections built on understanding and compassion. Readers are encouraged to apply these insights in their own lives, fostering awareness of the interconnected nature of their relationships and using this understanding to promote greater harmony and mutual support.

Entanglement and the Synchronization of Human Behaviors

In the fascinating realm of quantum physics, entanglement describes how particles become linked, remaining connected over any distance. This concept mirrors the remarkable synchronization seen in human behaviors, especially among close-knit groups and relationships. Individuals in such bonds often seem to tune into each other's rhythms, instinctively reacting and even anticipating each other's thoughts or emotions, much like entangled particles reflecting each other's state. These connections manifest subtly in everyday interactions, such as finishing each other's sentences or understanding without words. This challenges the view of people as isolated, proposing a more interconnected perspective that honors the invisible bonds uniting us.

Recent research has begun shedding light on this synchronization, blending insights from psychology, neuroscience, and quantum theory. A key element in this phenomenon involves mirror neurons, which activate when we perform an action or observe someone else doing it. These neurons act like connectors, enabling shared experiences that transcend individual boundaries. For example, couples who've been together for long periods often show synchronized heartbeats or brainwaves, highlighting their deep connection. This biological mirroring not only fosters empathy but also reinforces the symbiotic nature of relationships, echoing principles from quantum entanglement.

Viewing human behavior through this lens reveals that such interconnectedness extends beyond personal ties to larger social structures. In team settings, this connection can appear as a shared alignment of goals and actions, enhancing collaboration and productivity. Synchronization goes beyond shared tasks to include a collective mindset that allows for adaptive and fluid interactions. By nurturing these connections, organizations can build a more cohesive and resilient workforce, similar to how entangled particles contribute to the stability of quantum systems.

The applications of this understanding are vast, especially in conflict resolution and community building. Recognizing the interconnected nature of human interactions allows mediators and leaders to create environments that encourage synchronization, leading to better communication and problem-solving. Techniques like active listening and empathy training can enhance this natural inclination, helping individuals and groups align their perspectives. This approach not only resolves conflicts but also fosters stronger, more empathetic communities, reflecting the cooperative dynamics seen in entangled systems.

To deepen our understanding of behavioral synchronization, imagine a world where social technology, inspired by quantum principles, helps us visualize and optimize these connections. Such tools could map interpersonal dynamics, offering insights into how relationships shape behavior and decisions. By harnessing this knowledge, people could navigate social landscapes with greater awareness, fostering connections that are not only reactive but also proactively nurturing. This vision highlights the transformative potential of viewing social dynamics through a quantum lens, inviting us to embrace the interconnected nature of our shared human experience.

Exploring Non-Locality in Social Connections

Quantum entanglement, a concept that challenges traditional views, offers a unique way to understand social connections over distances. At its essence, this phenomenon suggests that once particles are connected, they remain linked regardless of how far apart they are. This idea reflects the invisible ties between people who feel in sync emotionally and mentally despite being physically apart. Recent studies indicate that non-locality, a key aspect of quantum physics, might appear in human relationships through the inexplicable closeness felt with distant friends or loved ones. Such connections go beyond geographical borders, hinting at a deep-seated network of human interaction similar to quantum systems.

Exploring this further, consider how non-locality might reshape our understanding of social cohesion. In an era where digital communication allows for instant interaction, the similarities with quantum entanglement become even more evident. Social exchanges, much like quantum particles, can maintain a form of harmony across great distances. While traditional psychology often credits shared history or emotional closeness for these phenomena, a quantum lens offers another view: a fundamental interconnectedness that might be universal. This shift in perspective encourages us to rethink how we view relationships, highlighting the potential of unseen forces in sustaining our social fabric.

The use of artificial intelligence in exploring non-locality sheds additional light on these social intricacies. AI systems, analyzing vast amounts of data on human interactions, can uncover patterns and connections that are not immediately obvious. These technologies enable the simulation of scenarios where interconnected relationships affect decisions and behavior, offering insights into the invisible bonds that unite us. The merging of AI and quantum theory within social sciences opens a new chapter where conventional understanding is challenged, paving the way for novel approaches to relationship dynamics.

Imagine people in various parts of the world simultaneously experiencing similar emotions or thoughts. While some might dismiss this as mere coincidence, quantum non-locality presents a compelling alternative explanation. The shared experiences during global events, where emotions seem to spread swiftly through societies, can be seen as evidence of this principle. These occurrences invite us to explore the nature of social influence and suggest that our grasp of human connections may only be a glimpse into a more complex system.

For those interested in applying these insights practically, recognizing non-locality's potential can lead to more empathetic and harmonious interactions. By accepting the possibility of unseen connections, individuals and organizations can create environments that prioritize emotional and mental harmony. Encouraging open communication and shared experiences can fortify these ties, allowing people to access the collective consciousness that non-locality implies. As we delve into the quantum foundations of social behavior, the opportunities for personal and societal growth appear as vast as the universe itself.

Implications of Quantum Entanglement for Conflict Resolution

In the field of conflict resolution, the principle of quantum entanglement offers unique insights that go beyond traditional methods. This concept from quantum physics suggests that particles become so interlinked that a change in one instantaneously affects the other, no matter the distance between them. This idea can serve as a powerful analogy for human relationships, highlighting the deep emotional, psychological, and social connections between individuals. Recognizing the interconnected nature of these relationships during conflicts can shift the focus from opposing viewpoints to a more comprehensive understanding of shared experiences and mutual influence. This perspective encourages considering not only one's own needs and perspectives but also those of others, promoting a more empathetic and collaborative approach to resolving disputes.

Recent advancements in AI and machine learning have allowed researchers to draw effective parallels between quantum entanglement and human interactions. By analyzing extensive datasets of human behavior, AI can uncover patterns resembling quantum connections, offering insights into how people and groups are intertwined. This data-driven method helps identify underlying issues in conflicts that might not be immediately visible, leading to more focused and effective resolution strategies. For example, AI can reveal the emotional and social bonds influencing a conflict, enabling mediators to address these aspects and foster understanding and reconciliation. This integration of AI not only enhances the resolution process but also highlights technology's potential to augment human insight in complex social dynamics.

Examining the non-locality aspect of quantum entanglement shows how conflicts often extend beyond the immediate parties involved, affected by broader social networks and cultural contexts. Recognizing this interconnectedness is vital for effective conflict resolution. For instance, workplace disputes may be influenced by organizational culture or external societal pressures, and addressing these factors can help craft solutions that tackle the root causes of conflict. By adopting a quantum perspective, practitioners can move beyond superficial solutions, exploring the systemic issues that sustain discord. This comprehensive approach not only resolves the immediate conflict but also contributes to long-term harmony and collaboration within communities.

Thought-provoking questions arise from considering the implications of quantum entanglement for conflict resolution. How might understanding the interconnected nature of relationships alter our approach to disagreements? What role do subconscious and non-verbal cues play in these interconnected interactions, and how can they be used to foster resolution? By encouraging reflection on these questions, we can inspire more innovative and effective conflict resolution methods. This reflection can lead to new methodologies that integrate quantum principles, deepening our understanding of the intricate web of human interactions and the potential for transformative change.

Actionable steps can be drawn from these insights to enhance conflict resolution practices. Encouraging open communication and active listening can help parties recognize their interconnections and the influence they exert on each other. Using AI-driven tools to analyze communication and behavior patterns can provide mediators with valuable insights to guide the resolution process. Additionally, creating environments that promote empathy and understanding can help individuals embrace the interconnected nature of their relationships, leading to more sustainable and positive outcomes. By applying these insights, individuals and organizations can develop more nuanced and effective conflict resolution strategies, paving the way for a more harmonious and interconnected society.

Superposition and Social Ambiguity: Navigating Multiple States of Being

Consider the intricate nature of identity within the social realm. Similar to a subatomic particle that exists in various states until it is observed, humans often traverse a world filled with ambiguity, adopting multiple roles and personas at once. Whether it's the professional identity at work or the laid-back demeanor among friends, individuals continually transition between different modes of being. This multiplicity, echoing the overlapping states of particle physics, is not just an intriguing phenomenon—it is a fundamental aspect of human life. It highlights the fluidity and adaptability needed to flourish in a world where expectations and perceptions are ever-changing. By drawing parallels between quantum concepts and social ambiguity, we set out to explore how this duality influences our interactions and choices.

Throughout this chapter, we will dissect the complexities of social roles and the inherent contradictions they entail, akin to a particle's simultaneous possibilities. As we delve deeper, the impact of uncertainty on decision-making becomes clear, shedding light on the hidden layers of choice that shape our lives. Our exploration goes beyond the surface, diving into the intricate dance of complexity within interpersonal relationships, where the richness of human connection is found. Each section acts as a stepping stone, constructing a framework for understanding how, much like in particle physics, uncertainty and potential coexist in the social sphere. Through this lens, we aim to embrace the multifaceted nature of human identity and its profound effect on our social dynamics.

The Duality of Identity in Social Contexts

In the complex landscape of social interactions, the idea of overlapping identities provides an intriguing way to explore how people embody multiple personas. Much like particles that exist in several states at once, individuals adapt their identities based on their social environments. This multifaceted nature highlights the intricate balance between personal identity and the roles assumed in various settings. Modern social psychology research emphasizes how identity is not a fixed construct but a dynamic spectrum that shifts with context and societal demands. This adaptability enhances social flexibility and resilience, allowing for nuanced responses to life's complexities.

In diverse or multicultural settings, the duality of identity becomes even more evident as people adjust behaviors and attitudes to align with different social norms. This adaptability fosters richer interactions and deeper connections, as individuals find shared understanding amidst diversity. Yet, this requires balancing authenticity with embracing various identity facets. The challenge of maintaining authenticity while adapting can drive personal growth, encouraging exploration and integration of different self-aspects meaningfully.

As individuals navigate social roles, they continuously negotiate between conflicting expectations and responsibilities. Balancing roles like parent, friend, or mentor involves prioritizing based on situational needs and personal values. Successfully managing these contradictions is a hallmark of social intelligence, enabling effective relationship management and decision-making aligned with broader life goals. Understanding this dual nature empowers individuals to embrace complexity without conforming to rigid identities.

The concept of overlapping identities also significantly impacts decision-making, where ambiguity and uncertainty often arise. Instead of seeing decisions as binary, recognizing a range of potential outcomes encourages exploration and innovation. This mindset fosters creative problem-solving and adaptive thinking, equipping individuals to thrive amid change. By acknowledging the numerous possibilities within any decision, choices become both informed and imaginative.

To leverage the power of overlapping identities, individuals can cultivate self-awareness and mindfulness. Practices like journaling or discussing with trusted friends can uncover the values and beliefs shaping one's identity. Promoting openness and curiosity allows exploration of identity's duality without fear, fostering more authentic social experiences. Embracing identity's dual nature leads to more meaningful interactions, encouraging genuine understanding and empathy in engagements with others.

Navigating Contradictory Social Roles and Expectations

In human society, social roles and expectations mirror the complexity of quantum systems, where entities can exist in several states at once until observed. Individuals often juggle conflicting roles, such as being both a leader and a team collaborator or balancing parenting with professional duties. Each role demands distinct behaviors. Recent studies in social psychology indicate that those who skillfully handle these overlapping roles develop better adaptability, enhancing their success in both personal and professional domains. By embracing the intricacies of these roles, one gains a more nuanced perspective, accommodating diverse experiences and viewpoints.

Navigating these roles requires a heightened awareness and flexibility similar to a quantum system's ability to occupy multiple states. This adaptability is crucial in today's rapidly changing world, characterized by interconnectivity and diverse environments. A study in the Journal of Personality and Social Psychology reveals that individuals who adopt a "both-and" mindset, as opposed to the "either-or" approach, manage role conflicts more effectively. This mindset fosters integration rather than compartmentalization of different roles, leading to harmonious interactions and reduced stress. Embracing this complexity allows for authentic self-expression and deeper connections with others.

The ambiguity of managing conflicting roles can significantly impact decision-making. When roles overlap or clash, individuals may experience cognitive dissonance, a psychological tension from holding contradictory beliefs or roles. This tension can spur creative problem-solving, encouraging exploration of innovative solutions that respect multiple aspects of one's identity. A recent meta-analysis in cognitive psychology highlights the value of this ambiguity, suggesting it leads to more flexible thinking and enhanced decision-making. By embracing these contradictions, individuals can discover new avenues for growth and innovation.

Interpersonal relationships thrive on complexity and unpredictability, akin to quantum systems. Recognizing the multifaceted nature of social roles can lead to more resilient and dynamic connections. When individuals accept the ambiguity of their roles, they engage in open, honest communication, fostering trust and understanding. This aligns with findings from a 2022 study on relational dynamics, which showed that couples who acknowledge and discuss role conflicts report higher satisfaction and intimacy. Viewing relationships through this lens allows appreciation of the fluid nature of human connections, resulting in richer interactions.

Practical strategies for managing these contradictory roles include cultivating mindfulness and self-reflection, which help individuals attune to their internal states and external demands. Regularly evaluating one's roles and expectations can identify areas of tension and suggest ways to integrate them harmoniously. Engaging in dialogue with others about role expectations fosters understanding and cooperation. Encouragingly, organizations are recognizing the benefits of supporting employees in balancing their roles through flexible work arrangements and open communication cultures. As society evolves, the ability to navigate conflicting roles and expectations will remain a valuable skill, empowering individuals to thrive in an increasingly complex world.

The Influence of Ambiguity in Decision-Making Processes

In the domain of decision-making, uncertainty often dominates, resembling the overlapping states seen in quantum physics. Imagine a scenario where each decision holds numerous potential outcomes, each influencing the others in a complex interplay of possibilities. This state of overlapping choices introduces a rich complexity in human decisions, as individuals weigh uncertainties and navigate through a landscape of potential results. Uncertainty becomes a compelling force that drives people to adapt and refine their approaches, similar to scientists modifying their models to address the inherent uncertainties in subatomic physics.

Consider a company poised to launch a new product in an unpredictable market. Executives confront a multitude of variables, such as consumer preferences, competitor maneuvers, and regulatory changes. Each element, akin to a subatomic particle, remains indeterminate until a decision is made. Studies in behavioral economics suggest that embracing this uncertainty, rather than avoiding it, can lead to more innovative solutions. Recognizing the coexistence of multiple potential scenarios allows decision-makers to prepare for a wider range of possibilities, enhancing strategic flexibility.

The influence of uncertainty in decision-making extends beyond the corporate sphere into personal choices, where the balance between known and unknown factors shapes everyday decisions. For instance, when selecting a career path, individuals often deal with incomplete information about future job stability, personal satisfaction, and financial prospects. This uncertainty mirrors the probabilistic nature of particle systems, where entities exist in several states until measured. By adopting a mindset that welcomes uncertainty, individuals can explore numerous pathways, enabling them to pivot and adapt as new information arises, leading to more informed and resilient life decisions.

In academia, researchers probe into the cognitive mechanisms that dictate how humans manage ambiguity in decision-making. Cutting-edge studies in neuroscience reveal that the brain's prefrontal cortex plays a vital role in processing uncertain information, much like a quantum computer managing complex algorithms. This understanding paves the way for developing decision-making models that better mimic human cognition, providing insights into how uncertainty can be leveraged rather than feared. By integrating these findings with AI, scholars are creating advanced tools capable of forecasting human behavior in uncertain environments, offering a glimpse into a future where ambiguity is not just a hurdle but an opportunity for advancement.

To effectively navigate the complexities of uncertainty in decision-making, individuals and organizations can adopt several strategies. One approach is to foster a culture of experimentation, where trial and error are valued as learning opportunities. This mirrors the scientific method, where hypotheses are tested against uncertainty. Additionally, encouraging diverse perspectives can reveal hidden possibilities, similar to observing a quantum system from various angles. By cultivating an environment that values flexibility, creativity, and open-mindedness, uncertainty can shift from a source of anxiety to a wellspring of potential, empowering individuals and organizations to succeed amid the unknown.

Embracing Complexity in Interpersonal Relationships

In interpersonal relationships, human interactions can be as intricate as quantum superposition. Just as particles exist in multiple states until observed, people often embody various roles, emotions, and identities within their social circles. This complexity requires a deep understanding of the subtle dynamics at play in our personal connections. Think of someone who juggles being both a caregiver and a professional; they navigate these roles with an inherent fluidity shaped by their experiences. This adaptability enriches relationships, fostering deeper connections when it is understood and embraced.

The idea of superposition in social contexts helps explain how individuals may hold conflicting desires and motivations. In relationships, this can appear as the simultaneous need for independence and closeness—a duality that, while challenging, can also be enriching. This mindset encourages acceptance and even celebration of contradictions, allowing for more genuine engagement with others. Recent psychological studies highlight the benefits of embracing such complexity, suggesting that individuals who integrate seemingly contradictory impulses often experience greater emotional intelligence and satisfaction in relationships. This perspective encourages moving away from binary thinking and adopting a more holistic view of human behavior.

Navigating the web of interpersonal relationships requires managing ambiguity and uncertainty, akin to the observer effect in quantum physics. By becoming aware of the multiple layers within themselves and others, individuals can better appreciate the unpredictable nature of human interactions. This awareness fosters resilience and adaptability, crucial traits for healthy relationships. By welcoming complexity, individuals can cultivate empathy and patience, essential in dealing with the ambiguity of human emotions and intentions. This approach aligns with social neuroscience research, which underscores the importance of mental flexibility in building strong relational bonds.

Embracing complexity in personal connections also prompts a reevaluation of traditional conflict resolution strategies. Conventional methods often seek clear-cut solutions, but a perspective inspired by quantum theory advocates for approaches that recognize the intricate nature of human experiences. This might involve understanding one's multifaceted needs and learning to communicate them effectively. By acknowledging the multiple states of being within a relationship, individuals can work towards solutions that respect these complexities rather than simplifying them. These insights lead to more sustainable and fulfilling interactions, inviting individuals to approach conflicts with nuance and depth.

Applying these principles in everyday life involves practices that foster an appreciation of complexity, such as mindfulness and reflective listening. These practices help individuals recognize and value the diverse aspects of their identities and those of others. Open dialogue about the various roles and states we inhabit can lead to more genuine and meaningful connections. By questioning assumptions and remaining open to the complex tapestry of human interactions, individuals can cultivate relationships that are both resilient and profoundly enriching. This approach challenges the status quo, inviting a deeper exploration of what it means to connect with others in an unpredictable and varied world.

Reflecting on our journey through the intriguing world of quantum mechanics and its unexpected links to social dynamics, we uncover a fresh layer of understanding. The core ideas of particle physics, such as interconnectedness and overlapping states, transcend their scientific origins to become potent symbols of human relationships and the nuances of social complexity. Interconnection underscores the deep bonds among individuals, indicating that our actions resonate beyond simple interactions, weaving complex webs of influence and compassion. Similarly, overlapping states encourage us to accept the diverse facets of our identities and decisions, fostering a richer appreciation for the various roles we inhabit across different scenarios. These insights pave the way for a transformative view of social interactions, where uncertainty and possibility intertwine to shape our shared experiences. As we venture further into the confluence of subatomic principles and societal behaviors, this exploration promises to illuminate the extensive ways in which the microscopic realm echoes the intricacies of human existence. Consider how these ideas might reshape your perception of social dynamics, challenging your assumptions and inspiring a more interconnected understanding of the world.

Chapter Two

Quantum Entanglement And Interpersonal Relationships

In the intricate dance of science and emotion, a captivating parallel emerges, challenging our perception of human bonds: quantum entanglement. Envision two particles, seemingly separate and distant, yet mysteriously intertwined, each reacting instantaneously to the other. This enigmatic connection mirrors the profound ties within our personal relationships, where moments arise that defy explanation—when the thoughts or feelings of someone dear resonate within us, regardless of distance. These connections hint at a reality where our relationships, much like entangled particles, operate beyond the physical, touching a realm where the invisible becomes tangible.

As we delve into this chapter, we embrace the idea that relationships are more than straightforward links; they are complex networks woven with shared experiences and emotional resonance. The science behind these deep connections becomes evident in instances when two people, bound by a shared understanding, seem to function as one. This phenomenon invites us to explore how our thoughts, emotions, and energies align, creating interactions that echo the harmony found in the quantum realm. By unraveling these connections, we gain insights into the subtle choreography of human interaction.

Yet, just as particles may lose their connection, human relationships can also experience dissonance. Quantum decoherence provides a lens to examine the breakdowns in communication and understanding that happen in our social interactions. Observing these disconnections reveals the delicate balance needed to sustain meaningful relationships. This exploration highlights the complexity of our social bonds and underscores the potential transformation that comes from adopting a quantum perspective on the ebb and flow of human connections. As we journey deeper into this chapter, the interconnected nature of our lives becomes a fascinating tapestry, interwoven with the threads of quantum theory and the fabric of human emotion.

How Relationships Mirror Quantum Entanglement: Instantaneous Connections

In a world where the boundaries of science are ever-expanding, the intriguing parallels between quantum physics and human relationships invite us to see connections in a new light. Picture two individuals, separated by vast distances, yet bound by an unseen connection that bridges their emotions and thoughts instantaneously. This phenomenon echoes the concept of quantum entanglement, suggesting that our relationships might defy physical space, much like the enigmatic nature of non-locality in the quantum realm. By examining these invisible links, we gain a richer understanding of the forces that quietly sustain our deepest bonds.

As we delve into this exploration, the harmony of emotions and thoughts over long distances becomes a focal point. How is it that some people can sense each other's feelings without a word, as if they share a mental frequency? This synchronicity points to a hidden depth in human interactions, akin to the intertwined states that characterize quantum phenomena. In the following subtopics, we will explore how these connections shape empathy and intuition, offering a fresh lens on the enduring enigma of human bonds. Moreover, we will look into how principles from quantum communication might enhance and fortify the ties that connect us. Through this perspective, the topic unfolds, promising to deepen our appreciation of the unseen dynamics at play in our relationships.

The Role of Non-Locality in Emotional Bonds

Quantum entanglement, where particles become intertwined and the state of one instantly affects the other, serves as a powerful metaphor for emotional bonds in human relationships. This idea of non-locality—where actions on one particle have immediate effects regardless of distance—parallels the deep connections found in interpersonal dynamics. Picture two people who, despite being miles apart, seem to sense each other's emotions or thoughts. This non-locality in emotional bonds indicates a form of instantaneous communication that transcends physical boundaries, much like entangled particles. The notion that emotions or moods can resonate across vast distances challenges traditional views of relationship dynamics and invites exploration into how such bonds might be nurtured or understood through a quantum perspective.

The alignment of thoughts and emotions across distances, similar to quantum entanglement, isn't just theoretical. Recent neuroscience and psychology studies suggest that deeply connected individuals often experience simultaneous emotional shifts, even when physically apart. This phenomenon, known as emotional contagion, can be observed among close friends, family members, or romantic partners. Advanced research methods, like fMRI scans and EEG monitoring, are beginning to unravel the neural pathways that might facilitate this mental alignment. These scientific discoveries suggest a biological foundation for such connections, providing a tangible basis for what might otherwise be considered abstract or mystical.

Entangled relationships, much like their quantum counterparts, can influence the development of empathy and intuition between connected individuals. The ability to intuitively sense another's feelings or anticipate their needs, often without explicit communication, showcases the depth of these connections. This intuitive understanding, which can seem almost telepathic, might be better comprehended through the paradigm of quantum entanglement. Recent research in social neuroscience indicates that mirror neurons play a crucial role in this process. These neurons, which activate both when we perform an action and when we observe someone else performing the same action, may serve as a biological foundation for empathy, reinforcing the concept of entangled minds.

Quantum communication models, which explore how information is shared and understood between entangled particles, can provide innovative frameworks for strengthening human connections. While traditional communication relies heavily on verbal and non-verbal cues, quantum models suggest the potential for more profound, intrinsic exchanges of information. By examining how particles communicate instantaneously, researchers are inspired to explore new ways humans might connect on deeper, more instinctual levels. Some emerging psychological practices and therapies aim to harness this potential, encouraging individuals to cultivate heightened awareness and presence with their partners, potentially leading to more resilient and fulfilling relationships.

For those interested in applying these quantum insights to their own lives, fostering emotional alignment and harmony can begin with simple, mindful practices. Encouraging open and empathetic dialogue, practicing active listening, and fostering a sense of shared presence can enhance the synchronicity and depth of personal bonds. Additionally, engaging in activities that promote emotional resonance, such as meditation or joint creative endeavors, can further cultivate these quantum-like connections. By adopting a quantum perspective on relationships, individuals might not only deepen their understanding of their own bonds but also contribute to a broader societal shift toward more empathetic and interconnected communities.

Synchronization of Thoughts and Feelings Across Distances

In quantum physics, entanglement offers an intriguing analogy to human relationships, highlighting how people can experience a deep connection regardless of physical distance. Just as entangled particles are linked despite being separate, human bonds often transcend physical space, forming a complex web of emotional and cognitive links. This isn't just poetic; it has a basis in science. Research in interpersonal neurobiology shows that close relationships can lead to a mirroring effect, where thoughts, emotions, and even physical responses align between individuals. These studies suggest an invisible thread that binds us, facilitating an exchange of feelings and ideas beyond conventional communication.

A closer look at this phenomenon reveals the influence of mirror neurons, which play a crucial role in empathy and shared experiences. These neurons activate not only when performing an action but also when observing someone else perform it, establishing a neural foundation for empathy. The emotional and cognitive resonance between people mirrors the quantum entanglement of particles, where a change in one instantly affects the other. This neurological harmony is evident in shared laughter among friends or the silent understanding between long-time partners. Emotional alignment can even extend to physiological synchronization, such as matching heart rates during profound connections, underscoring the depth of these ties.

Quantum communication models offer a fascinating framework for understanding and enhancing these connections. In the quantum realm, information can be exchanged instantly between linked particles, providing a template for more intuitive, non-verbal human interactions. This perspective invites us to cultivate and harness subtle communication modes in our daily lives. By being attuned to unspoken cues and emotional signals from those we care about, we can nurture stronger, more resilient relationships. This concept challenges the conventional reliance on words, encouraging exploration of a deeper, more instinctual form of connection.

The implications of this emotional linkage span into intuition and empathy, where people report heightened understanding and anticipation of others' needs and feelings. Such experiences resemble a form of emotional telepathy, where thoughts and emotions are exchanged without words. This intuitive bond is especially strong in relationships characterized by trust and openness, suggesting that the quality of our interactions can significantly influence the depth of our connections. The idea that our emotional states can intertwine so deeply offers a new perspective on empathy's power and the potential for nurturing more meaningful relationships.

To apply this quantum-inspired understanding practically, fostering environments that promote emotional attunement and openness is essential. Practices like mindfulness and active listening enhance our sensitivity to others' emotional states, enabling more empathetic and understanding responses. Embracing the notion of emotional harmony allows individuals to cultivate a heightened awareness of interconnectedness, leading to more harmonious and fulfilling interactions. This approach not only enriches personal relationships but also has the potential to transform broader social dynamics, promoting a more compassionate and interconnected society.

The Influence of Quantum Entanglement on Empathy and Intuition

Quantum entanglement, a fascinating concept where particles remain connected regardless of distance, offers a fresh perspective on empathy and intuition in human relationships. This viewpoint suggests that even when separated by physical space, people often share deep emotional and intuitive links. Such interconnections imply that emotional ties might operate similarly to quantum entanglement, where one person's state can instantly affect another's, beyond physical limits. This idea challenges traditional views of emotional bonds, proposing that the essence of human relationships could be woven with the non-local threads of emotional resonance.

Consider the remarkable alignment of thoughts and emotions often seen among close friends or family members. Situations where one partner seems to understand the needs or feelings of the other without direct communication might be more than coincidence. Recent research in psychology and neuroscience indicates this phenomenon might resemble quantum entanglement, with mental and emotional states becoming intertwined. This complex synchronization could be seen as an intuitive dance, not just a result of shared experiences but potentially a sign of a deeper, quantum-like connection that enhances empathy. These bonds create an environment where intuition flourishes, enabling individuals to sense the emotional climate of their relationships, even from afar.

The convergence of quantum mechanics and human interaction invites exploration into the mechanisms behind empathetic and intuitive experiences. Quantum communication models, typically applied to particles, are now being examined for their relevance in understanding human connectivity. This perspective suggests that just as particles can exchange information without direct contact, humans might engage in a form of non-verbal, non-local communication. This prompts intriguing questions about empathy's nature: Is it merely a psychological construct, or could it involve a more intricate, possibly quantum, interaction between people? Addressing these questions could transform our understanding of empathy, shifting it from an emotional response to a phenomenon with a potential scientific and quantum basis.

Integrating quantum insights into the study of empathy and intuition expands our understanding and offers practical applications. Recognizing the possible quantum nature of these connections allows individuals to create environments that nurture empathetic and intuitive bonds. Practices like mindfulness and meditation could heighten sensitivity to these non-local links, fostering deeper, more meaningful relationships. Additionally, AI and machine learning models could map and predict patterns of empathy and intuition, providing insights into how these connections form and evolve. This knowledge empowers individuals to navigate their social landscapes more effectively, leveraging intuitive abilities to build stronger, more resilient relationships.

Reflecting on the quantum dimensions of empathy and intuition encourages a reevaluation of human interaction. It challenges us to consider that our social ties are not just biological or psychological but may be deeply intertwined with the universe's fundamental principles. Embracing this perspective opens the door to a new understanding, where empathy is not merely an emotional response but a sophisticated interplay of quantum-like interactions. As we explore these frontiers, we are reminded of the profound complexity and beauty in human relationships, encouraging us to engage with the world and each other in more enriched and meaningful ways.

Quantum Communication Models in Strengthening Human Connections

In our modern world, where human interactions extend beyond physical limitations, quantum communication offers a revolutionary approach to forging deeper connections. Inspired by the mysterious realm of quantum mechanics, these models provide a way to overcome traditional communication obstacles. By exploring entanglement and superposition, we uncover how people, even when far apart, can maintain meaningful bonds that feel instant and profound. Recent research suggests that the essence of such quantum-like interaction lies in shared emotions and synchronized thoughts, which are hallmarks of our most significant relationships. This perspective encourages us to investigate how unseen strands of empathy and understanding can form a network of strong, lasting connections.

The idea of "telepathic" communication between closely connected individuals emerges as an intriguing possibility when exploring quantum communication. Although not literal telepathy, this interaction resembles the quantum phenomenon where entangled particles influence each other regardless of distance. Cutting-edge studies in neuroscience and psychology propose that similar mechanisms might explain how close individuals seem to intuitively sense each other's thoughts or feelings. Imagine two people, separated by miles, who can perceive each other's emotional states or predict actions without spoken words. Such occurrences hint at a type of connection beyond conventional explanations, urging us to adopt a deeper, quantum-inspired view of human relationships.

Applying quantum principles to communication prompts us to reconsider the dynamics of empathy and intuition. For example, advanced research in quantum biology suggests that living beings might use quantum effects to process information, indicating a biological foundation for such quantum-like interaction. This idea invites a radical rethinking of empathy, viewing it not just as a learned ability but as an intrinsic part of human interaction, potentially embedded in our very biology. By embracing this viewpoint, we can better understand why some people have a remarkable ability to comprehend others, a talent that could be nurtured through intentional practices inspired by quantum communication models.

Practical insights from these models suggest strategies to nurture and deepen these connections. Creating environments that promote openness and vulnerability can enhance the quantum-like effects of intertwined relationships. Encouraging practices like mindfulness and active listening helps cultivate the mental states needed for aligning thoughts and emotions, strengthening the implicit channels of communication between individuals. These strategies not only improve personal relationships but also have significant implications for organizations and societies, where better communication and understanding can lead to more cohesive and harmonious interactions.

Envisioning the application of these quantum communication insights in everyday life, one might consider how adopting a quantum mindset could transform relationships. How might interactions change if individuals approached each conversation with the aim of creating an "entangled" connection, focusing on mutual understanding and respect? By exploring these possibilities, readers are encouraged to experiment with integrating these models into their lives, fostering a world where human relationships are not only maintained but deeply enriched, surpassing the limitations of time and space.

Emotional Synchronization: The Science Behind Deep Bonds

Picture two people in a bustling room whose eyes meet across the expanse, and suddenly, an unspoken understanding bridges the space between them. This phenomenon, often described as being in sync, mirrors the intriguing dance of quantum entanglement, where particles are linked in ways that challenge traditional explanations. Emotional resonance, the invisible thread weaving through the tapestry of human connection, echoes the mysterious harmonies found in the quantum world. As we delve into the complexities of this profound bond, we begin to understand how deeply our emotions can intertwine with those around us, beyond mere words.

Consider the dance of neural resonance, akin to a quantum symphony where alignment fosters empathy and insight. By exploring the role of mirror neurons, we uncover how they orchestrate a ballet of shared emotions, reflecting and intensifying feelings effortlessly. Then, contemplate the notion of interconnected minds, where interactions defy physical boundaries, prompting us to rethink our views on relationships. Each of these elements offers a window into the science behind deep emotional ties, inviting us to reconsider the nature of our interpersonal bonds through a quantum perspective. Here, the lines between individual minds blur, revealing a complex web of shared consciousness and emotional unity.

Neural Resonance and Emotional Connectivity

Neural resonance is a captivating phenomenon that offers insights into how people form profound connections. This concept, akin to musical instruments harmonizing in an orchestra, illustrates how our brains synchronize during significant interactions. Studies utilizing functional magnetic resonance imaging (fMRI) show that when individuals share experiences or engage in focused dialogue, their brain activity often mirrors each other. This alignment creates a sense of understanding and closeness, much like a symphony's harmonious blend of melodies. These findings highlight the intricate dance of the human brain in forging connections, indicating that our neural pathways are finely tuned to resonate with those we cherish.

Delving into neural resonance encourages us to explore the crucial role brain waves play in emotional bonds. Oscillatory brain activity, particularly within gamma and theta frequencies, is associated with heightened empathy and shared emotional experiences. Research suggests that when two people are emotionally aligned, their brainwaves display a coherent pattern, facilitating deeper connections. This neural coherence not only strengthens interpersonal ties but also enhances our capacity for empathy and understanding. The implications are vast, suggesting that nurturing environments that encourage this resonance can lead to more meaningful and resilient relationships.

The science of neural resonance reaches beyond emotional connectivity, prompting us to contemplate its applications across various fields. In education, for instance, teachers who achieve brainwave synchronization with their students may enhance learning outcomes by fostering an atmosphere of shared understanding and engagement. Similarly, in therapeutic settings, clinicians who align with their clients on a neural level might facilitate more effective interventions. These insights push us to appreciate neural resonance as a tool for improving human interaction, one that can be leveraged to boost communication and collaboration across different environments.

While the concept of neural resonance forms a robust framework for understanding emotional connectivity, it's vital to consider diverse perspectives within this field. Some researchers propose that this phenomenon might extend beyond face-to-face interactions, occurring even in digital communication where physical presence is absent. Others suggest that neural resonance could be amplified by practices like mindfulness or meditation, which are known to enhance awareness and empathy. This variety of viewpoints enriches our understanding, underscoring the need for ongoing exploration and dialogue within the scientific community to uncover the full potential of neural resonance.

Recognizing the transformative power of neural resonance invites us to reflect on how we can apply these insights to daily life. Simple practices like active listening, maintaining eye contact, and being present in conversations can cultivate the neural alignment that strengthens bonds. By prioritizing these elements in our interactions, we can deepen our emotional ties, creating a ripple effect that extends to broader societal dynamics. Much like the quantum realm, the world of human interactions is filled with unseen forces that profoundly shape our experiences and relationships.

Quantum Coherence in Empathy and Understanding

Quantum coherence offers a fascinating perspective on the complexities of empathy and understanding in human interactions. This phenomenon, where particles behave like waves and maintain a phase relationship, parallels the profound bonds shared by individuals who deeply empathize with each other. In the emotional realm, this coherence appears as a harmonious alignment of feelings and thoughts, akin to an unseen thread connecting people. Just as quantum coherence allows particles to exist in multiple states at once, human empathy enables individuals to grasp and appreciate diverse viewpoints, fostering richer connections.

Neuroscience has uncovered intriguing parallels between quantum coherence and neural synchronization during empathetic interactions. Using functional MRI and EEG, studies show that when individuals empathize, their brain wave patterns exhibit remarkable harmony, echoing the coherence seen in quantum systems. This neural resonance not only enhances emotional comprehension but also creates a shared experience that transcends speech. Within this synchronized state, individuals access a deeper level of connection, where words fall secondary to the unspoken language of empathy.

A crucial component of this phenomenon is coherent communication's role in sustaining empathy and understanding. Social interactions thrive on coherence, requiring openness and receptivity to allow emotions and thoughts to flow freely. This openness reflects the principle of quantum superposition, where multiple possibilities coexist until observed. Similarly, empathy flourishes when preconceived notions are set aside, embracing complexity and ambiguity. This approach not only strengthens bonds but also fosters mutual respect and understanding.

In practical terms, embracing quantum coherence in empathy can transform communication strategies and interpersonal relationships. Individuals can enhance empathy by practicing active listening and being attuned to non-verbal signals, aligning their emotional states with others. By creating environments that invite authentic expression and vulnerability, people can cultivate spaces where empathy and understanding thrive. This approach enriches personal relationships and has broader implications for collaboration in professional and community settings.

To fully harness the potential of quantum coherence in enhancing empathy, challenging conventional viewpoints and exploring unconventional methods is essential. Encouraging reflection on personal biases and engaging with diverse perspectives can lead to a deeper understanding of others' experiences. By embracing complexity and interconnectedness in human interactions, individuals can nurture a more empathetic society, where emotional coherence resonates across social boundaries. This paradigm shift, inspired by quantum principles, invites a reconsideration of relational dynamics and the transformative power of emotional coherence.

The Role of Mirror Neurons in Synchronizing Emotions

In the early 1990s, the discovery of mirror neurons profoundly transformed our understanding of empathy and emotional connection. Located in the premotor cortex and other related brain regions, these neurons activate both when we perform an action and when we observe another doing the same. This dual activation highlights a biological basis for shared experiences, enabling individuals to naturally perceive the emotions and intentions of others. Thus, mirror neurons act as a neurological bridge, fostering an intuitive connection between people that extends beyond mere observation into shared emotional territories.

The impact of mirror neuron activity goes beyond simple mimicry. When we see others experiencing emotions like joy or sorrow, our mirror neurons reflect these states, offering a biological foundation for empathy. This suggests that emotional alignment is not merely learned but is an intrinsic part of human nature. Functional MRI studies demonstrate that similar neural circuits are engaged whether we directly experience an emotion or witness it in others, emphasizing the shared pathways that aid in understanding and emotional harmony.

Exploring further, consider mirror neurons' role in social bonding and group unity. These neurons aid in developing rapport and mutual comprehension, allowing people to synchronize their emotions and actions effectively. In team environments, robust mirror neuron activity can enhance cooperation and communication, as individuals become more adept at anticipating each other's needs and reactions. This neural mirroring is crucial for successful collaboration, where emotional and cognitive alignment are vital.

However, the influence of mirror neurons raises questions about individuality and authenticity in social interactions. While emotional alignment fosters empathy and connection, it also poses the risk of conforming to group emotions at the cost of personal authenticity. Understanding this dynamic opens avenues to explore how individuals can maintain their unique emotional identities while benefiting from the empathy and connectivity mirror neurons provide. It invites reflection on balancing emotional resonance with self-expression in social contexts.

As research continues to uncover the complexities of mirror neurons, potential applications emerge in fields like conflict resolution, education, and mental health. By leveraging these neurons, strategies can be developed to enhance empathy and understanding across various settings, improving relationships and reducing social tension. The concept of mirror neurons not only sheds light on emotional alignment but also offers practical insights into nurturing deeper connections in an increasingly fragmented world. Thus, they represent a fascinating intersection of biology and social dynamics, offering a glimpse into the profound interconnectedness of human experience.

Entangled Minds: Non-Local Interactions in Human Bonds

In the sphere of human relationships, the notion of intertwined minds encourages us to ponder the unseen links bridging people across distances and various situations. Similar to the phenomenon of quantum linkage, where particles stay connected irrespective of the space between them, human bonds can display deep connections that surpass mere physical closeness. These remote interactions often reveal themselves in the remarkable ability of partners, family, or close friends to perceive each other's emotions or thoughts even when separated. Such occurrences challenge our conventional views of communication, suggesting that emotional and mental ties might function on a more complex and intangible level. Emerging research in neuroscience and psychology supports this idea, showing that when people form profound emotional bonds, their brain activity can synchronize in ways reminiscent of quantum coherence.

Delving further into these phenomena, we encounter the fascinating field of quantum cognition, which applies quantum theories to explain human thought processes. This view proposes that minds in close relationships might operate in a state of entanglement, where shared experiences and emotions weave a tapestry of linked thoughts. The implications are significant, hinting that our mental states are not isolated but part of a broader network of shared consciousness. This interconnectedness might clarify why individuals in strong relationships often develop an intuitive grasp of each other's needs and feelings, seemingly predicting actions or emotions before they're spoken. The principles of quantum superposition play a role here, as multiple potential outcomes coexist until one becomes reality, much like the myriad possibilities in human communication and understanding.

Examining this further, the importance of remote interactions becomes evident in the realm of empathy and social intuition. Quantum models suggest that empathy might not solely be a learned behavior or mirror neuron activity but could stem from an inherent intertwining of human consciousness. This perspective redefines empathy as an intrinsic capability, potentially enhanced by quantum coherence, where individuals in close emotional proximity experience a merging of mental states, leading to profound mutual understanding. Practically, this could transform how we approach conflict resolution, communication, and emotional support, highlighting the significance of nurturing these invisible bonds to foster healthier, more understanding relationships.

The study of intertwined minds also prompts us to consider the role of technology and AI in understanding and enhancing these connections. With AI advancements, we are beginning to see tools that can analyze and even predict human emotions and interactions by examining patterns not easily discernible. This opens exciting possibilities for enriching our social interactions, offering insights into how we can strengthen our ties with others. For example, AI-driven platforms might one day assist individuals in cultivating deeper relationships by identifying and reinforcing the quantum-like bonds that unite us, leading to more harmonious and fulfilling interactions.

In essence, the idea of intertwined minds challenges us to rethink the boundaries of human interaction and consciousness. By embracing the concept that our thoughts and emotions are linked in ways akin to quantum particles, we open new avenues for understanding and improving our relationships. This perspective not only enhances our appreciation for the complexity of human bonds but also inspires us to explore the untapped potential of our collective consciousness. As we continue to unravel the mysteries of entanglement in both physics and psychology, we are reminded of the profound interconnectedness that defines our social world, urging us to nurture and cherish these intangible yet powerful connections.

Breaking Entanglments: How Social Disconnects Reflect Quantum Decoherence

How do our cherished bonds unravel when the seamless dance of human interactions encounters turbulence? The intricate choreography of relationships mirrors the delicate balance of quantum states. Just as quantum entanglement weaves particles together in a mysterious embrace, our connections often reflect this phenomenon, forming links that stretch beyond the ordinary confines of time and space. However, when these ties are tested or broken, it resembles the process of quantum decoherence. This unraveling of bonds, akin to the loss of coherence in quantum systems, compels us to explore the enigmatic forces that lead to social disconnections. By investigating these parallels, we gain a deeper insight into how the threads of our social fabric can fray and, in some cases, completely disengage. This journey not only illuminates the dynamics of personal relationships but also offers a lens through which larger societal structures can be understood.

Throughout this exploration, we will uncover the intriguing role of quantum decoherence in the dissolution of human relationships, drawing parallels between the collapse of quantum states and the alienation often experienced in social contexts. Consider how social disturbances, much like environmental noise in quantum systems, can gradually erode the connections we hold dear. We will also delve into how the quantum framework can provide insights into the disbandment of groups, offering a novel perspective on the dynamics of collective disengagement. This exploration aims to illuminate the subtle yet profound ways in which the principles of the quantum realm can shed light on the complexities of human interaction, guiding us toward a deeper appreciation of the delicate equilibrium that sustains our social ties.

The Role of Quantum Decoherence in Human Relationship Breakdowns

In the intricate interplay of human relationships, the concept of quantum decoherence provides an intriguing framework for understanding the gradual disintegration of interpersonal bonds. Decoherence in quantum physics refers to the process where a quantum system loses its coherent properties, transitioning from a superposition to a single state due to environmental interactions. This transformation can be compared to the breakdown of a relationship, where once seamless interactions fragment under external pressures or unresolved internal issues. In both relationships and quantum systems, maintaining coherence is crucial for sustaining mutual understanding and shared experiences. When coherence diminishes, misunderstandings and emotional distance often follow.

Consider how external 'noise' affects both quantum systems and human relationships. In physics, noise disrupts the coherence of a quantum state. Similarly, in relationships, stressors like work pressure, societal expectations, or personal challenges can introduce discord, weakening emotional bonds. Vibrant communication may become strained, and shared experiences that once unified individuals may lose their significance. This analogy highlights the importance of creating an environment that fosters relational coherence, where open communication and mutual support act as safeguards against disruptive noise threatening the partnership.

Recent research draws parallels between quantum decoherence and the psychological processes contributing to relationship breakdowns. Social psychology studies suggest that unresolved conflicts and unmet needs can lead to emotional decoherence, where partners disconnect from each other's emotional realities. This mirrors the quantum state collapse, where potential outcomes narrow to a singular, often undesirable state. By recognizing these patterns, individuals can address underlying issues proactively, preventing irreparable disconnection. Effective conflict resolution and regular emotional check-ins can sustain relational coherence, ensuring partners remain attuned to each other's needs and perspectives.

When exploring decoherence in group dynamics, similar parallels arise. Just as a quantum system loses coherence when isolated from its environment, a social group can fragment when its members become isolated from one another. This social decoherence can manifest as communication breakdowns, loss of shared vision, or lack of collective engagement. To counteract this, fostering an inclusive environment is crucial, where each member feels valued and connected to the group's objectives. Encouraging participation and open dialogue helps maintain the group's coherence, akin to how quantum systems retain coherence when shielded from disruptions.

Applying these insights practically, individuals and groups can adopt a proactive approach to maintaining relational coherence. This might involve setting aside regular time for open communication, engaging in collaborative problem-solving, and creating spaces where each voice is heard and respected. Reflecting on past interactions and identifying patterns of disconnect can offer valuable insights into areas needing attention and improvement. By understanding decoherence dynamics in personal and group settings, individuals can foster more resilient and coherent connections, enriching their social interactions and experiences. This quantum-inspired perspective not only deepens our understanding of human relationships but also provides innovative strategies to navigate the complexities of social life.

Analogies Between Quantum State Collapse and Social Isolation

The phenomenon of quantum state collapse, where a quantum system shifts from a superposition of states to a single state upon observation, serves as an intriguing metaphor for understanding social isolation. Similar to how a quantum system loses its many possibilities, individuals facing social isolation may experience a reduction in their social options. This can lead to feeling 'stuck' in a solitary existence, lacking the vibrant interactions that define a healthy social life. The analogy between quantum state collapse and social isolation is not just conceptual; it offers insights into how people withdraw from community life, resulting in a sense of disconnection.

Current research increasingly focuses on the psychological effects of social isolation, drawing parallels to the impact of quantum collapse on particles. Studies suggest that prolonged isolation can cause cognitive decline and emotional strain, much like how a particle's potential diminishes once its state collapses. This comparison can inspire novel interventions to reintegrate isolated individuals into social networks. By viewing isolation through the lens of quantum physics, we can devise strategies that restore individuals to a state of 'superposition', allowing them to reengage in diverse social roles and opportunities.

The idea of social 'noise'—the myriad influences and distractions in one's environment—can also be compared to factors causing quantum state collapse. Just as environmental interference can lead a quantum state to collapse, social noise can overwhelm a person, resulting in withdrawal and isolation. This perspective encourages us to consider how spaces could be designed to minimize harmful noise and foster positive interactions, creating environments where individuals can maintain a balance of engagement without risking social withdrawal. By cultivating surroundings that reduce negative social interference, we can strengthen the durability of interpersonal bonds.

The analogy further extends to examining how group dynamics mimic quantum behaviors. In a group, individual isolation can lead to the 'collapse' of group cohesion, similar to how a single particle's collapse can affect a quantum system. This viewpoint challenges organizations and communities to identify signs of social isolation within their networks. By addressing factors that lead to individual isolation, groups can maintain their collective unity, ensuring all members remain involved and connected. This proactive stance can prevent the breakdown of group dynamics, akin to sustaining stability in a quantum system.

This exploration raises an important question: How can principles of quantum mechanics inform strategies to combat social isolation? By applying quantum insights, we can create innovative methods to reconnect isolated individuals. Programs that emulate the unpredictable yet interconnected nature of quantum systems—where individuals can freely transition between various social roles and relationships—may offer pathways to reintegration. This approach not only deepens our understanding of social isolation but also provides actionable steps for nurturing inclusive communities that thrive on diversity and interconnectedness.

Exploring the Effects of Social Noise on Relationship Entanglement

In the realm of human relationships, social noise represents the multitude of external influences that can disrupt the delicate equilibrium between individuals, similar to how quantum interference affects linked particles. This social noise includes societal pressures, cultural norms, digital distractions, personal stresses, and communication breakdowns. These elements can create a complex symphony of challenges that threaten the stability of emotional bonds. Viewing social noise through the lens of quantum decoherence offers insight into how these disturbances gradually weaken connections that were once strong and intertwined, leading to a metaphorical 'collapse' of harmonious relationship states.

Quantum decoherence occurs when a quantum system loses its coherence due to environmental interactions, resulting in the loss of its unique quantum characteristics. Similarly, in human interactions, the introduction of social noise can disrupt emotional and psychological alignment between people. Research in social psychology shows that negative influences, such as persistent misunderstandings or societal stigmas, can erode trust and compassion, much like an environment causing decoherence in a quantum system. By recognizing these parallels, individuals and communities can develop strategies to identify and reduce the impact of social noise, thereby preserving the coherence of their relationships.

Imagine a couple navigating the stress of modern life, bombarded by digital notifications and societal expectations. This constant social noise can lead to misaligned priorities, causing partners to fall out of sync with each other's emotional wavelengths. Just as quantum particles need isolation from environmental interactions to maintain coherence, relationships benefit from creating spaces for intentional communication and mutual understanding, protecting them from the chaotic outside world. Practical steps like digital detoxes, mindfulness practices, and open dialogues can serve as protective measures, fostering a more coherent and resilient interpersonal connection.

The concept of social noise also extends to larger group dynamics and societal interactions. In organizations or communities, collective coherence can be disrupted by conflicting interests, cultural biases, or misinformation. By applying quantum principles, leaders and group members can adopt frameworks that prioritize transparency and shared objectives, reducing the decohering effects of social noise. Studies in organizational behavior suggest that encouraging diverse perspectives and open communication helps maintain group coherence, much like quantum systems shielded from decoherence by controlled environments.

Encouraging reflection on personal experiences, one might ask: How does social noise appear in your daily interactions, and what steps can you take to lessen its impact? By adopting a quantum mindset, individuals can become aware of external influences that threaten their connections. Engaging in introspective practices and fostering environments of trust and openness can enhance relationship coherence, allowing individuals to navigate the complexities of modern life with renewed understanding and resilience. Exploring social noise and its effects on relationship entanglement invites deeper contemplation of how we can harmonize with those around us, despite the ever-present background hum of the world.

Quantum Decoherence as a Model for Understanding Group Disbandment

Quantum decoherence offers a fascinating lens through which to explore the breakdown of group unity. Like entangled particles, individuals in a group move together toward shared goals, demonstrating cohesion through mutual understanding. Yet, when disruptions arise—be it through conflicting aims, external pressures, or miscommunication—this unity begins to erode. This process is akin to quantum decoherence, where environmental interactions disrupt a quantum state, leading to loss of entanglement. Recognizing these parallels enhances our understanding of factors behind group disbandment and underscores the fragile balance needed to sustain unity.

In group dynamics, quantum decoherence serves as a metaphor for how minor disturbances can fracture a once-coherent system. Take, for instance, a close-knit team collaborating on a project; each member's contribution is crucial, much like particles in a coherent state. When differences in vision or direction surface, similar to external noise in a quantum setting, these minor issues can grow, undermining collaboration. This highlights the need to spot and address such disturbances early, promoting open dialogue and flexibility to preserve group coherence.

Social noise mirrors environmental interactions in quantum decoherence, with external influences like rumors or societal changes disrupting internal harmony. In social contexts, noise manifests in various forms, gradually eroding cohesion. To counteract this, groups can adopt strategies similar to error correction in quantum systems, such as enhancing internal communication and reinforcing shared values. These measures can mitigate external noise's impact, maintaining the group's unity and focus.

By drawing analogies between quantum decoherence and group disbandment, we gain insights into mechanisms that can prevent or delay such outcomes. Encouraging diverse perspectives within groups boosts adaptability, akin to quantum systems benefiting from a superposition of states. This diversity helps groups explore multiple solutions, reducing stagnation and building resilience against challenges. Moreover, fostering an environment where feedback is valued and applied can act as a stabilizing force, maintaining coherence amid potential disruptions.

This quantum perspective on group dynamics not only provides a metaphorical framework but also inspires practical methods to maintain group integrity. Encouraging critical thinking through questions like "What external factors might be causing disruptions?" or "How can we integrate diverging opinions into a coherent strategy?" can lead to proactive measures. Understanding and applying these principles helps individuals and organizations navigate complex group interactions, ensuring coherence and unity even in adversity.

Examining the concept of quantum entanglement in the realm of human connections uncovers a striking parallel between the intricate world of subatomic particles and the complexities of our relationships. Much like entangled particles, our bonds can defy physical separation, emphasizing the enigmatic yet undeniable links that unite us with others. This delicate interplay of emotional harmony reflects the profound depth of our connections, mirroring the intricate interactions observed in nature. Yet, just as quantum decoherence highlights the transient nature of particle connections, moments of social disconnect remind us of the fragile and ever-changing nature of human ties. By understanding this dynamic, we gain a valuable perspective on the ebb and flow of our relationships, urging a conscious effort to nurture them while remaining open to change. As we delve deeper into the similarities between quantum principles and social dynamics, we are encouraged to reconsider our notions of interconnectedness, potentially reshaping our understanding of relationships and promoting more meaningful interactions. This journey into the quantum dimension challenges us to rethink the essence of our bonds, paving the way for further exploration of how such principles influence broader societal patterns.

Chapter Three

Superposition And Decision Making In Social Contexts

In life's quieter moments, when you're at a crossroads deciding whether to call an old friend or keep working, speak up in a meeting or stay silent, explore a new opportunity or stick to your comfort zone, you're engaging with a concept as old as time: the idea that multiple possibilities exist until you choose. This notion, echoed in quantum physics, suggests that until we make a decision, various outcomes coexist. It reflects the intricacies of human choices, where our potential actions are suspended like Schrödinger's cat, both alive and dead until observed.

Consider how each of us navigates through the overlapping roles we play every day. In one circle, you might be the joker, spreading joy and laughter. In another, you might be the thoughtful thinker, sharing insights and wisdom. These simultaneous identities influence our interactions and shape our social environments. This chapter explores how these roles coexist without settling into one fixed identity until necessary, highlighting the natural intricacies of our social lives. Our ability to embrace these different states offers a more nuanced understanding of human behavior and decision-making.

As we explore the idea of coexistence in social settings, we encounter moments when choices transform from mere possibilities into concrete actions. These instances, where potential turns into reality, provide insights into our decisions and the frameworks of societal norms and expectations. By examining these phenomena, we gain a deeper understanding of how possibility and reality interact in our social worlds, encouraging us to reflect on the identities we shape and the choices that define us.

The Role of Superposition in Everyday Choices: Multiple Possibilities

Imagine waking up each morning, surrounded by a tapestry of choices, each shimmering with potential like stars in a vast universe. From the instant our eyes open, we find ourselves in a world brimming with possibilities, where every decision we contemplate can lead to a unique reality. This idea of multiple outcomes coexisting mirrors the concept of coexistence in quantum physics, where particles exist in various states until an action determines their fate. Our choices, much like these particles, exist in a realm of potential until we decide to act. This dance of possibilities is not merely a theoretical exercise; it is a concrete aspect of our daily experiences, influencing both mundane and significant choices.

In the complex web of our social lives, the choices we make are further shaped by the environments and groups we engage with. Each social situation presents its own array of potential outcomes, requiring us to navigate various roles and identities, akin to the overlapping states in quantum mechanics. Whether pondering personal decisions or participating in intricate group dynamics, viewing our choices through this lens reveals the layered and often unpredictable nature of human behavior. This chapter delves into how our decisions, initially suspended in potential, eventually crystallize into action, much like the resolution of a quantum state. By examining decision-making from this perspective, we gain a deeper understanding of the intricacies of human interaction and set the stage for a more profound exploration of our social world.

Exploring Superposition in Personal Decision Trees

In the field of personal decision-making, examining choices through the lens of superposition unveils their intricate nature. Similar to subatomic particles that exist in various states at once, individuals face decisions that embody multiple potential outcomes. Imagine having to decide between a stable career and an entrepreneurial path. Initially, both choices exist simultaneously in the mind, each leading down a unique path filled with possibilities and uncertainties. This mirrors quantum superposition, where multiple outcomes are possible until a choice is made, akin to a quantum state solidifying upon observation.

The interaction between personal choices and social contexts adds layers of complexity. Social norms, cultural expectations, and peer influences act as external forces, shaping the perceived value and feasibility of different options.

Consider a student selecting a college major. This decision isn't made in isolation; family expectations, societal trends, and peer influences converge, creating a web of influences that contribute to the overlapping state of potential academic paths. This dynamic highlights the importance of recognizing how external factors impact decision-making, similar to how environmental variables affect particle behavior in quantum systems.

Group decision-making introduces further intricacies, as individual choices intertwine with the collective will. In collaborative environments like business teams or community groups, each person brings their own array of potential choices, creating a complex network of intertwined possibilities. The challenge is to navigate these layered options to reach a consensus, much like finding coherence in a quantum system. This process often involves negotiation, compromise, and aligning personal aspirations with group goals, illustrating the balance between individual agency and collective dynamics.

Adopting a quantum perspective on decision-making under uncertainty invites a re-evaluation of traditional approaches to assessing risks and benefits. Instead of relying solely on linear models of probability, embracing the inherent uncertainty of superposition offers a more comprehensive understanding of decision-making processes. This perspective encourages openness to multiple possibilities, fostering adaptability and resilience in the face of uncertainty. By recognizing the potential for various outcomes to coexist, decision-makers can better prepare for a range of scenarios, enhancing their ability to navigate complex and unpredictable situations.

To apply insights from superposition in practical scenarios, individuals can adopt strategies that promote flexibility and openness to change. Engaging in scenario planning or adopting an experimental mindset allows exploration of diverse paths without prematurely settling on a single choice. By maintaining an awareness of this overlapping state, individuals cultivate a deeper understanding of their options, leading to more informed and intentional decisions. This approach aligns with contemporary research on cognitive flexibility, empowering individuals to make choices that reflect their values and adapt to changing circumstances. Through this quantum-inspired lens, decision-making becomes a dynamic process of exploration and discovery, rich with potential and opportunity.

The Influence of Social Contexts on Choice Superpositions

In the world of decision-making, the concept of overlap provides an intriguing way to explore the complex nature of human choices. Borrowed from quantum mechanics, this idea suggests that a person's selections remain unfixed until they are acted upon or observed. Within social contexts, this translates to the many possibilities that exist before someone commits to a particular path. For example, when contemplating a career, an individual might initially consider a variety of roles—such as teacher, engineer, or artist—each representing a potential future that exists simultaneously until a definitive choice is made. The complex web of options is shaped by personal values, aspirations, and external influences, all contributing to a dynamic landscape of decision-making.

The role of social environments is crucial in shaping these overlapping choices. People often navigate varying expectations and norms depending on the social setting, which can significantly alter the range of potential selections. For instance, one might behave differently when making decisions within a family context compared to a professional one, each providing a unique set of options. This fluidity resembles how electrons can exist in multiple states until measured. A person may balance multiple personas, like being a supportive friend and a competitive colleague, with each context influencing their decision-making process. This ability to maintain such overlapping states enables individuals to adapt and succeed in diverse social settings, reflecting the intricate interaction between personal agency and social impact.

Recent studies in psychology and sociology highlight the significant influence of social contexts on overlapping choices. Research indicates that group dynamics can either broaden or narrow the choices available to individuals. In collaborative settings, diverse perspectives can lead to a rich overlap of ideas, fostering innovative solutions that may not emerge in isolation. Conversely, the pressure to conform can limit this overlap, reducing the range of possibilities by discouraging dissent. Understanding these dynamics provides valuable insights into decision formation and optimizing processes within collaborative efforts.

Quantum-inspired models in social sciences suggest a promising role for artificial intelligence in predicting and analyzing these overlapping choices. By simulating various social contexts and potential outcomes, AI can offer enhanced tools for navigating complex decision-making environments. Such technology can provide predictive insights that recognize the fluid nature of human choices, enabling more informed strategies. Imagine AI systems aiding in career counseling by modeling potential career paths based on an individual's interests, social influences, and market trends, thereby empowering them to make more informed choices.

To apply these insights practically, individuals and organizations can adopt strategies that embrace the multiplicity of options inherent in overlapping choices. Promoting open-mindedness and creating environments where diverse perspectives are valued can expand the range of possibilities considered in any decision-making process. Increasing awareness of how social contexts influence choices can lead to more conscious and deliberate decision-making. By recognizing the fluid and context-dependent nature of choices, individuals can navigate their social worlds more effectively, unlocking new possibilities and achieving desired outcomes.

Superposition and the Complexity of Group Decision-Making

In the sphere of collaborative decision-making, the concept of overlap emerges as a blend of various potential outcomes, each influenced by the distinct preferences, viewpoints, and backgrounds of the participants involved. This complexity is reminiscent of the quantum principle where particles exist in multiple states until measurement. In a similar fashion, groups navigate a landscape filled with possibilities, with each member contributing to the collective decision-making framework. This dynamic exchange of ideas can lead to original solutions, as the fusion of diverse perspectives may yield results unattainable in isolation. Engaging with this idea highlights the richness that diversity brings to decision-making, emphasizing the importance of inclusive dialogue in multifaceted social situations.

Imagine a team tasked with selecting a strategic path for a project. Each member contributes their own priorities and ideas, creating an overlap of potential directions. This environment can foster creative energy, where these combined possibilities culminate in an outcome that surpasses individual efforts. Nonetheless, such overlap also poses challenges, demanding effective communication and negotiation skills to navigate complexity and reach consensus. Viewing overlap as an opportunity encourages teams to welcome ambiguity and use it as a springboard for innovation, rather than perceiving it as a hindrance to decision-making.

Research in organizational behavior indicates that successful group decision-making often relies on balancing individual freedom with group unity. This finely-tuned equilibrium resembles achieving quantum coherence, where differing elements align harmoniously while maintaining their distinct identities. By nurturing an environment where team members feel confident in expressing their ideas while staying open to others', groups can maximize their decision-making potential. Techniques such as structured brainstorming and choice matrix analysis can help preserve this balance, ensuring that diverse thoughts enhance rather than impede the decision-making process.

The role of technology, particularly artificial intelligence, in aiding group decision-making is becoming increasingly significant. AI can process large volumes of data and detect patterns that might escape human observation, providing insights into group dynamics. By simulating various decision-making scenarios, AI assists groups in foreseeing the potential outcomes of different choices, effectively narrowing the range of possibilities to those most advantageous. This technological progress highlights the importance of integrating AI as a tool to improve collective decision-making, while still valuing the unique human traits of empathy and intuition.

As groups tackle the intricacies of decision-making, they must also deal with the unpredictability inherent in human behavior. This unpredictability, reminiscent of the quantum uncertainty principle, implies that despite sophisticated models and analyses, some aspects of group dynamics will remain elusive. Embracing this uncertainty can be empowering, encouraging groups to stay adaptable and open to change. By fostering a mindset that values flexibility and continuous learning, teams can transform uncertainty from a source of anxiety into a catalyst for opportunity, promoting a culture of resilience and innovation. Through this perspective, the quantum approach not only deepens our understanding of group decision-making but also inspires practical strategies for thriving in an increasingly complex world.

Quantum Perspectives on Decision-Making Under Uncertainty

In the realm of uncertain decision-making, the notion of coexistence provides a profound perspective on human experiences. This idea suggests that multiple potential outcomes can exist simultaneously until a definitive decision is reached, much like the numerous possibilities individuals encounter when faced with complex choices. This challenges the conventional linear approach, proposing instead that people, akin to particles in a quantum state, navigate a range of possibilities before settling on a specific action. By understanding choice as a dynamic process where numerous alternatives are considered at once, individuals can become more adept at handling uncertainty, embracing the fluid nature of decision-making rather than being hindered by indecision.

In social settings, this concept highlights the intricate network of factors that influence decisions. Social interactions, cultural norms, and personal biases all contribute to the state of coexistence where decisions reside. This multidimensional view encourages a deeper understanding of how social contexts shape the range of choices available. In a group, for instance, the complexity of decision-making can increase, with each member's personal coexistence interacting with others, creating a collective decision landscape rich with possibilities. Understanding this complex interaction can enhance collaboration, allowing groups to leverage diverse thoughts and reach more innovative solutions.

Recent advancements in artificial intelligence and machine learning offer new ways to understand decision-making amid uncertainty. These technologies can simulate and analyze scenarios, providing insights into the probabilistic nature of human decisions. AI algorithms, parallel to quantum computing's ability to process multiple states simultaneously, offer a novel approach to modeling and predicting decision patterns. By incorporating AI into decision frameworks, individuals and organizations can gain a clearer picture of various potential outcomes, enabling more informed and strategic decision-making. This technological integration offers practical tools for navigating the complexities of modern life, where uncertainty is ever-present.

Amidst scientific and technological advancements, it's essential to consider the philosophical implications of viewing decision-making through this lens. This perspective encourages a shift from deterministic thinking, embracing ambiguity and uncertainty as inherent human conditions. By recognizing that multiple truths can coexist and that decisions occur within a fluid and ever-changing landscape, individuals can develop a mindset that is more adaptable and resilient. This philosophical shift has the potential to transform how people approach choices, fostering a culture open to new possibilities and exploring uncharted paths.

To apply these insights practically, individuals can start by acknowledging the coexistence inherent in their choices. Embracing uncertainty as a natural part of decision-making, rather than something to fear, empowers people to take calculated risks and explore diverse options. By consciously considering multiple scenarios and outcomes, they can make more holistic and informed decisions. Cultivating environments that encourage dialogue and diverse perspectives can help groups avoid decision-making stalemates and harness the collective wisdom within their varied experiences. Through these actions, the principles of quantum-like decision-making can become part of daily life, offering a nuanced and effective way to navigate the complexities of human interaction.

Social Superposition: Maintaining Multiple Identities in Different Groups

Picture yourself navigating a life filled with diverse roles, each as distinct as characters in a theatrical performance. One moment, you're the committed professional, driven by team objectives, and the next, you're a compassionate friend, sharing a coffee and a listening ear. These roles coexist within you, much like overlapping waves, each waiting for its moment to emerge based on context and companionship. This fluid interplay forms the essence of human interaction, where we deftly shift and adjust our personas without losing our fundamental essence.

In this intricate dance of identities, group dynamics play a crucial role, influencing how we reveal ourselves. The expectations and norms within different social settings can either empower or restrict how we express our self-concept, much like an observer shaping the behavior of elements in a scientific experiment. Grasping this dynamic offers deep insights into the nuances of human behavior. As we journey through a rapidly changing world where technology, including AI, sheds light on our social patterns, we gain a chance to understand these complex interactions with clarity. By applying strategic insights, we can navigate social expectations while using AI to model and predict these identity shifts, offering a window into the rich tapestry of the human experience.

Navigating Identity Through Quantum Superposition in Social Roles

Understanding the complex nature of identity within social roles can be compared to the concept of quantum superposition, where one can exist in multiple states at once. In social settings, this means an individual can adopt various identities in different groups, each revealing unique aspects of their character and values. For example, someone may be a committed professional at work, a caring parent at home, and an adventurous companion among friends. This diversity promotes a flexible and adaptive way of engaging with society, fostering a richer understanding of oneself and others. Recent social psychology research indicates that this adaptability is not only natural but beneficial, enhancing resilience to stress and improving social connections. By embracing multiple identities, people can more effectively navigate the intricate social structures they encounter daily.

Advancements in quantum cognition offer further insights into how superposition can be applied to identity. These studies suggest that the brain processes information similarly to quantum processes, enabling the simultaneous consideration of multiple roles. This ability can be invaluable in a world where societal expectations are ever-changing, encouraging an openness to change and fostering greater empathy and understanding in various social contexts. Viewing identity through the quantum superposition lens allows individuals to appreciate the fluidity of their roles and make informed choices about which aspects of their identity to emphasize in different situations.

Despite the intriguing framework of social superposition, challenges arise. Managing multiple identities can lead to conflicts when expectations from different groups clash. Here, strategies inspired by quantum theory, such as maintaining coherence and navigating uncertainty, can provide innovative solutions. By applying these principles, individuals can achieve a harmonious balance, ensuring no single role overshadows others. This balance is crucial for mental health and fostering genuine interactions in diverse social environments. It also promotes a holistic view of identity, acknowledging the interconnectedness of various roles and their importance in shaping a well-rounded self-concept.

AI plays a crucial role in modeling these intricate identity patterns, offering insights previously unattainable. Advanced algorithms can analyze extensive data to identify subtle patterns in how individuals manage their identities across social groups. This can lead to personalized strategies tailored to each person's social environment. AI's capacity to process and learn from complex datasets provides a deeper understanding of the nuanced ways in which social roles are navigated, offering practical tools for optimizing social interactions. These technological advancements open new avenues for personal growth and social harmony.

As we delve into the idea of social superposition, the implications for personal development and societal progress are significant. By embracing identity fluidity and leveraging AI to deepen our understanding, individuals can cultivate a more adaptable and resilient approach to life. This perspective encourages continuous learning and growth, fostering a society that values diversity and inclusivity. Provocative questions emerge: How can we use our understanding of social superposition to enhance communication across cultural divides? What role does this concept play in driving innovation within organizations? By challenging traditional notions of identity and exploring new paradigms, we can unlock the potential for transformative change in both personal and collective realms.

The Impact of Group Dynamics on Individual Identity Expression

The complex interplay of personal identity within group dynamics mirrors the fascinating concept of quantum overlap. In social settings, people often present various personas, adjusting their identity expressions to meet the expectations of different groups. This multifaceted persona allows them to exist in multiple social "roles" at once, similar to particles in a state of overlap. This adaptability enables individuals to navigate intricate social environments with skill, ensuring their sense of belonging across diverse communities. As they change in response to social stimuli, they exhibit a unique resilience and flexibility, reflecting the inherent uncertainty of the quantum realm.

In exploring the psychology of group dynamics, researchers have uncovered how collective behaviors shape individual identity. A study by social psychologist Dr. Susan Cross reveals that individuals frequently alter their self-views to align with group norms, a process she calls "social convergence." This can lead to the temporary suppression of certain identity aspects while allowing more contextually suitable traits to surface. Group pressure, implicit bias, and the need for acceptance all play a role in this fluid identity expression. Yet, the overlap state permits individuals to revert or adapt these traits based on the group context, mirroring quantum phenomena.

AI technology is shedding light on these intricate identity patterns, offering new insights into social overlap. By examining vast datasets from social media and online interactions, AI systems can detect patterns in how people adjust their identities across various groups. These insights provide a deeper understanding of social conformity and personal authenticity, showing how digital personas often represent a spectrum of identity states. Advanced AI models, such as those from MIT's Media Lab, are beginning to predict changes in identity expression influenced by group dynamics, opening a window into the future of social interaction analysis.

Balancing multiple identities presents both challenges and opportunities in social contexts. Individuals frequently encounter conflicting expectations from different groups, requiring them to navigate these tensions adeptly. Quantum strategies, like probabilistic thinking and embracing uncertainty, can offer practical solutions. By adopting a mindset that values flexibility and open-ended possibilities, individuals can better manage the pressures of diverse social expectations. This approach not only promotes personal growth but also fosters a richer, more interconnected social experience.

Readers are encouraged to examine their social interactions through this quantum perspective. Reflect on situations where you adjust your identity to meet different group expectations. How can you use this adaptability to strengthen your social connections while staying true to your core self? Engaging with these reflections can lead to a deeper appreciation of the complex relationship between individuality and group dynamics. As society becomes more interconnected, understanding the parallels between quantum mechanics and social behavior offers profound insights into the evolving nature of human identity.

Balancing Conflicting Social Expectations with Quantum Strategies

Navigating conflicting social expectations can feel like moving through a maze of overlapping realities, akin to the principle where particles exist in multiple possibilities at once. This serves as a metaphor for how people juggle diverse expectations from various social circles. Each group comes with its own norms and demands, which may sometimes clash, creating tension for those trying to maintain balance. Adopting a strategic approach allows individuals to manage these expectations, shifting smoothly between roles and adopting the most fitting persona for each situation while preserving their core identity.

Recent studies highlight that humans have an inherent adaptability to handle different social roles, a skill that can be refined using concepts inspired by quantum mechanics. Just as particles do not settle into a state until observed, individuals can keep options open, delaying the 'collapse' of their social identity until the context necessitates it. This strategic ambiguity provides flexibility, aligning with various group norms without experiencing internal conflict. This method not only alleviates stress but also enhances one's ability to thrive in complex social environments, fostering a more resilient sense of self.

Similar to how particles maintain a harmonious state among disparate conditions, individuals can cultivate harmony in their social relationships. By identifying common values and goals across different groups, one can create a cohesive narrative that aligns with multiple personas without conflict. This requires an understanding of personal values and effective communication, ensuring core principles remain intact even as external expectations shift. Such coherence acts as a stabilizing force, allowing for seamless transitions across diverse social domains.

Innovative research in AI offers promising tools to model and anticipate complex social identity patterns. By analyzing vast datasets of social interactions, AI can reveal patterns and trends that might not be immediately apparent. These insights empower individuals to anticipate and navigate potential conflicts in social expectations, offering strategic pathways to harmonize diverse group demands. AI-driven simulations also provide hypothetical scenarios, allowing individuals to experiment with different approaches to maintaining social balance without real-world consequences.

The journey to mastering social coexistence is ongoing, requiring a willingness to embrace complexity and uncertainty. By adopting strategies inspired by quantum physics, individuals can transform the challenge of conflicting social expectations into opportunities for growth and innovation. This perspective encourages a nuanced understanding of identity, one that is dynamic and responsive to the ever-changing social landscape. It invites readers to reflect on their own social roles and consider how they might apply these strategies in their daily lives, fostering a more harmonious and adaptable social existence.

Leveraging AI to Model Complex Social Identity Patterns

In today's intricate social landscape, artificial intelligence stands as a guiding light in unraveling the complexities of human identity. By harnessing AI to explore the concept of social coexistence, we can envision a future where individuals manage multiple roles across varied social settings with remarkable clarity. AI's capabilities in data analysis and pattern recognition open opportunities to understand the complex interplay of overlapping social identities. As people engage in different social contexts—whether at work, with family, or among friends—they assume multiple personas, similar to particles existing in a state of coexistence. AI-driven models enable researchers to simulate interactions among these diverse identities, revealing how individuals adapt their roles to meet the expectations of different groups.

Further exploration reveals AI's potential to uncover hidden dynamics within social groups that influence personal identity expression. Machine learning algorithms can discern patterns in how individuals modify their self-presentation based on group interactions, offering insights into the mechanics of identity shifts. For example, AI can analyze social media activity to understand how individuals display various facets of their identities across online platforms. This data-centric approach highlights the adaptability of human identity, providing a framework for understanding how group dynamics can both constrain and liberate personal expression. By examining these patterns, AI becomes a tool not only for observation but also for creating environments where authenticity can flourish amid diverse social expectations.

AI also provides strategies for navigating conflicting social demands, promoting a balanced existence within the overlap of identities. Advanced modeling allows AI to simulate scenarios where individuals face opposing expectations from different groups, predicting outcomes and suggesting optimal resolutions. This predictive capability equips individuals with strategic foresight to manage social tensions, helping them maintain balance among their various identities. Additionally, AI can tailor these strategies by incorporating unique personal traits and past experiences, offering customized solutions for managing identity conflicts. In this role, AI serves as a guide, aiding individuals in reconciling diverse aspects of their lives while respecting the complexity of their social identities.

The potential of AI extends beyond observation into the realm of innovation, where it actively models complex identity patterns within societal frameworks. By utilizing neural networks and deep learning, AI can simulate large-scale social systems, providing insights into how individual identity overlaps contribute to broader societal trends. This comprehensive view allows researchers to foresee how changes in identity expressions might influence cultural norms or societal evolution. As AI refines its models with additional data, its predictive accuracy improves, offering a glimpse into the future of social dynamics. These insights can inform policymakers and social leaders, guiding the development of inclusive policies that embrace the multifaceted nature of human identity within society.

As AI continues to advance, its role in modeling identity patterns becomes increasingly important, challenging traditional notions of identity. By promoting a deeper understanding of social coexistence, AI paves the way for transformative societal changes. It encourages us to reconsider the nature of identity in a world where the boundaries between self and others blur, inviting us to embrace the diverse dimensions of our social existence. Through this perspective, AI not only reveals the intricacies of human identity but also inspires us to envision a future where our varied selves coexist harmoniously within the complex web of society. As we stand at the intersection of technology and humanity, AI's capacity to model and understand social coexistence holds the promise of a more interconnected and empathetic world.

Collapsing the Social Wave Function: When Decisions Solidify into Actions

One intriguing facet of decision-making within social contexts is how choices solidify when under observation. Much like a quantum wave function collapses when measured, our many potential paths converge into definitive actions when scrutinized by others. This dynamic underscores the relationship between personal autonomy and societal influence, highlighting that our choices are not just individual—they are also sculpted by our social environments. Making a decision becomes a dance between internal drives and external pressures, a nuanced choreography that, while often unseen, is deeply impactful. Navigating this balance, social observation acts as a powerful force, turning possibilities into reality. These processes highlight that human decisions are interconnected events, resonating through the social fabric and shaping, as well as being shaped by, the collective mindset.

Within this complex web of social interactions, the concept of indecision emerges as a potent force that can either destabilize or strengthen group cohesion. The tension between multiple options and the necessity of commitment often mirrors the uncertainty found in quantum systems. Delayed or unresolved decisions can destabilize social groups, leading to fluctuations in unity and purpose. However, once a choice is made, it can unify group identity and direction, similar to a collapsing wave function that defines a particle's trajectory. The dynamics of choice under social pressure shed light on the intricate balance of freedom and constraint, where external forces can both illuminate and obscure decision-making. In this context, AI technologies offer intriguing potential as observers in social decision processes, providing insights and predictions that can guide or even alter human interactions. By exploring these ideas, we gain a deeper insight into how decisions, often perceived as solitary acts, are intricately woven into the broader tapestry of social life.

The Role of Social Observation in Decision Formation

In the intriguing world of human decision-making, social observation significantly influences how choices are formed. Much like how particles in physics change behavior when watched, people often alter decisions under societal scrutiny. This observer effect suggests that decisions are made in a social context, influenced by the expectations and judgments of others. The presence of an observer, whether real or imagined, can shift decision-making from potential options to concrete actions.

Consider how social observation plays out in group settings like workplaces or communities. Here, peer expectations can steer individuals towards decisions that align with group values. Behavioral psychology studies show that people often conform to group norms, not just out of compliance but as a reaction to being observed. This conformity can promote group harmony and unity, making decision-making a collaborative process where social observation encourages prioritizing collective over personal interests. Thus, decisions reflect a shared social consciousness rather than individual thought alone.

The role of digital observers adds another layer of complexity in today's world. Social media platforms act as ever-present observers, influencing decisions on a large scale. The digital traces individuals leave invite feedback from a wide audience, often resulting in decisions seeking approval or validation. This digital observation can heighten conformity pressure but also allows individuals to present multiple identities, with each decision showcasing a different aspect of their persona. The pervasive nature of digital observation highlights the importance of understanding its impact on solidifying decisions.

The interplay between social observation and decision-making raises questions about personal autonomy and agency. To what extent are our decisions truly our own versus influenced by societal scrutiny? This invites introspection on personal decision-making processes. Recognizing social observation's influence empowers individuals to make choices with greater intention. By acknowledging this impact, one can develop strategies to ensure decisions align with personal values despite external pressures.

For those looking to apply this understanding, mindfulness and self-awareness are key. Identify situations where social observation might be affecting decisions. Reflect on motivations behind choices and consider if they align with personal goals or merely conform to social norms. Engage in open conversations with others to gain diverse perspectives and challenge the assumptions behind decisions. Embracing these practices turns the observer effect into a tool for informed and genuine decision-making, enriching personal growth and strengthening social structures.

Quantum Indecision and Its Impact on Group Cohesion

The intriguing concept of quantum indecision in social contexts draws parallels between the uncertain nature of quantum physics and the complexities of human interactions. Just as a quantum particle exists in overlapping states until observed, individuals often face numerous potential actions and outcomes, leading to indecision. This indecisive state resembles the condition of coexistence in quantum mechanics and can manifest in social groups as hesitation or ambivalence, impacting the group's ability to act cohesively. When members experience this indecision, the group struggles to reach consensus, delaying necessary decisions and actions.

Quantum indecision affects not only individual choices but also the broader dynamics of group cohesion. Successful groups rely on shared goals and synchronized actions, yet indecision can disrupt these processes, introducing uncertainty and fragmentation. Members might perceive indecision as a lack of leadership or direction, leading to diminished morale and engagement. Studies in social dynamics suggest that prolonged indecision can erode trust and weaken group bonds, underscoring the need for effective strategies to maintain unity. Understanding these dynamics involves recognizing how individual indecision aggregates to influence collective behavior, offering insights into fostering resilience and adaptability in social structures.

Advancements in AI provide promising solutions for addressing indecision within groups by acting as an impartial facilitator in decision-making processes. Analyzing communication patterns, AI can pinpoint moments of indecision and propose solutions. By simulating various outcomes, AI offers predictive insights, allowing groups to navigate choices confidently. Integrating AI into decision-making frameworks can mitigate indecision's paralyzing effects by presenting clear, evidence-based options aligned with group goals. This capability not only enhances group cohesion but also empowers individuals to overcome indecisive states, fostering a more dynamic and responsive social environment.

Social pressure significantly influences quantum indecision, either exacerbating or alleviating it. Strong social norms can compel individuals to conform, collapsing the multiplicity of possibilities into decisive actions. This pressure can act as a catalyst for decision-making, providing the impetus to move past indecision. However, if perceived as coercive, it may increase resistance, prolonging indecision. Cultivating an environment that balances encouragement with autonomy helps individuals and groups navigate social pressures effectively, facilitating smoother transitions from indecision to action.

Encouraging critical reflection on these phenomena invites individuals to consider their experiences of indecision and its effects on their social interactions. Questions like "How does indecision impact my role within a group?" or "How can AI tools assist in decision-making?" can lead to actionable insights. Recognizing the parallels between quantum physics and social behavior allows individuals to discover innovative strategies for overcoming indecision, enhancing their ability to contribute positively to group dynamics. Through the lens of quantum theory, individuals can gain a more nuanced understanding of decision-making processes, ultimately fostering stronger, more cohesive social networks.

The Dynamics of Choice Under Social Pressure

In social environments, choices are shaped by a web of influences where peer pressure and surroundings significantly impact individual decisions. The interaction between personal desires and outside expectations resembles a complex dance, akin to a quantum state where multiple possibilities coexist. Social pressure acts as a force, pushing individuals into a condition where potential actions overlap, each revealing different aspects of conformity or defiance. This coexistence of choices, swayed by others' presence, creates a multifaceted landscape where decisions are not solely personal but are shaped by the collective gaze.

Within this social framework, decision-making under stress unveils intriguing dynamics. Under peer scrutiny, the metaphorical 'social wave function' of potential choices tends to collapse into definitive actions. This mirrors quantum mechanics, where observation influences a particle's state. Social observation can magnify certain decisions while suppressing others, compelling individuals to align with perceived group norms. This societal expectation can steer individuals toward choices that may not reflect their true preferences, highlighting the influence of social observation in shaping decisions.

Recent studies into this dynamic reveal how group influence can strengthen or weaken cohesion. Indecision arises when social pressures create a conflict between the need for acceptance and the desire for authenticity. In such scenarios, individuals may face decision paralysis, where fear of making the wrong choice under scrutiny leads to inaction. This indecision can ripple through a group, affecting its momentum and potentially disrupting harmony. Studying these phenomena provides insights into how groups maintain stability or fall into discord, suggesting strategies to create environments that balance individual freedom with collective unity.

Artificial Intelligence offers a new perspective for examining these complex social dynamics. As an observer, AI can analyze decision-making patterns under pressure, providing a non-human viewpoint that identifies subtle cues and biases. Machine learning algorithms can process extensive social interaction datasets, offering predictions and insights to refine our understanding of decision-making in pressured settings. Integrating AI into social analysis not only uncovers hidden patterns but also proposes methods to reduce the negative effects of social pressure, fostering environments where authentic choices can thrive.

Navigating these complex social terrains requires strategies to enhance decision-making resilience. Emphasizing self-awareness and reflection helps individuals recognize external influences and realign decisions with personal values. Encouraging open group dialogues can alleviate social pressure, fostering a culture that values diverse opinions and reduces fear of judgment.

By understanding the dynamics of choice under social pressure and utilizing tools like AI, individuals and groups can create environments that embrace the positive aspects of social influence while preserving personal agency and integrity.

AI as an Observer in Social Decision Processes

Artificial intelligence has become a powerful force in social decision-making, functioning as a contemporary observer with the capacity to significantly influence outcomes. This role finds interesting echoes in quantum physics, where observation can change a system's state. In social settings, AI-equipped algorithms analyze extensive datasets, acting as observers that can steer decision paths by offering insights or predictions once out of reach. These technologies guide individuals and groups through intricate decision trees, presenting options and probabilities that might otherwise stay hidden. By revealing unseen patterns in social behaviors, AI not only aids decision-making but also plays a crucial role in turning potential choices into real actions.

AI's capacity as a social observer surpasses basic data analysis. It can explore the complex layers of human interactions, detecting subtle cues and nuances that might escape human notice. In scenarios with intricate social dynamics, AI can unravel influences, showing how individual choices merge into collective decisions. For instance, an AI system analyzing social media trends might shift public opinion or consumer behavior, effectively shaping current cultural moods. Offering unmatched clarity and foresight, AI empowers individuals and organizations to navigate social decision-making complexities with greater precision.

However, AI's role as an observer prompts crucial questions about agency and autonomy. While AI delivers valuable insights, a delicate balance exists between depending on algorithmic guidance and preserving human intuition and independence. When AI recommendations dominate decision-making, there's a risk of diminishing human creativity and spontaneity. This interplay between AI analysis and human agency encourages reflection on harnessing technology's strengths while preserving human judgment's intrinsic value. Thought-provoking scenarios arise when considering AI's influence in high-stakes environments like healthcare or finance, where data-driven insights and human expertise must converge.

To fully leverage AI as an observer in social decision processes, individuals and organizations can adopt strategic approaches. Fostering collaboration between AI systems and human decision-makers can create a synergistic environment where both contribute their strengths. Integrating AI-generated insights into brainstorming sessions can lead to innovative solutions neither AI nor humans might achieve alone. Additionally, ensuring transparency in AI algorithms helps decision-makers understand the rationale behind AI observations, enabling informed choices that integrate data-driven and human-centric perspectives. By embracing these strategies, the fusion of AI and human judgment can result in more nuanced and effective decision-making.

Imagining AI's future as a social observer invites speculation about its evolving role in shaping society. As AI advances, its ability to predict and influence social dynamics will likely grow, offering new possibilities for understanding and guiding human behavior. This evolution presents opportunities to explore ethical considerations, such as ensuring AI's role aligns with societal values and priorities. By contemplating these potential developments, readers engage with the ongoing dialogue about AI's place in social decision-making, considering how its capabilities might be harnessed to promote positive outcomes while respecting the complexity and diversity of human interactions.

Exploring the idea of overlap within social realms reveals the intricate ways individuals manage a landscape brimming with possibilities. Throughout this chapter, we've seen that, much like particles in quantum theory, people exist in states of potential, constantly balancing different personas and options until a situation demands a definitive choice. This viewpoint enhances our grasp of decision-making, highlighting adaptability and flexibility as vital elements of human conduct. By adopting this lens to view our social interactions, we realize that decisions are not just endpoints; they are fluid processes shaped by context and observation. Embracing this concept of social overlap encourages reflection on the numerous possibilities inherent in our everyday lives and interactions. This insight deepens our appreciation of personal agency and lays the groundwork for exploring how these ideas extend into larger societal frameworks. As we move forward, we should contemplate how this perspective can shape our understanding of collective decision-making and the evolution of social norms, setting the stage for an in-depth analysis of group behavior dynamics in the upcoming chapter.

Chapter Four

Observers And Observed The Quantum Observer Effect In Society

Envision a lively café brimming with energy on a hectic morning—a snapshot of society where each glance, nod, and silence crafts a rich mosaic of human interaction. Seated at a quiet corner table, you become an observer of this unfolding scene. Your silent presence subtly influences the atmosphere. The barista, sensing your gaze, adds a touch of flair to their routine. Conversations at nearby tables shift slightly as patrons become aware they are being watched. This phenomenon, where observation alters behavior, mirrors a captivating concept from physics: the observer effect.

As we journey through this chapter, we explore intriguing connections between the quantum realm and human society, both shaped by the act of observation. In particle physics, measuring a particle can alter its state, challenging our grasp of certainty and existence. Similarly, in our social world, being watched can shape our actions, decisions, and self-awareness. This interplay between the observer and the observed is not merely an interesting detail but a vital theme woven into the fabric of our lives, influencing everything from personal interactions to societal norms.

Through examining the impact of observation, we gain fresh insights into how our roles as spectators and participants shape the social environments we navigate. Each interaction becomes a dance, choreographed by the unseen eyes that watch and the conscious minds that adjust in response. The ripple effect of observation extends beyond individuals, influencing group dynamics and altering the flow of conversations and social norms. This chapter invites us to explore how awareness of being watched—or watching others—can transform our social landscape, offering a unique perspective on the intricate dance of human connection.

The Role of the Observer in Quantum Mechanics: A Brief Overview

At the crossroads of theory and practice lies the intriguing role of the observer in the realm of quantum physics—a concept that has captivated both scientists and philosophers for many years. This idea proposes that the act of observing can directly affect the behavior of particles, raising profound questions about reality and consciousness. This effect challenges the conventional belief in an objective universe, where events occur independently of observation, suggesting instead a universe that reacts to the presence and actions of those watching. This fascinating interaction between observer and observed prompts us to reflect on how our perceptions shape the world, not only in quantum phenomena but also in everyday life. As we delve into this concept, we begin to notice similarities with human interactions, where the presence of a viewer can subtly or significantly change behavior.

In this exploration, we trace the historical development of the observer's role within quantum theory, uncovering key experiments that have highlighted this phenomenon. These experiments, often yielding unexpected and paradoxical outcomes, have sparked ongoing debates about consciousness's role in shaping reality, stirring both scientific and philosophical communities. The implications extend beyond the confines of the laboratory, addressing fundamental questions about existence and our place in it. By examining how quantum physics challenges our perceptions, we gain insight into the social impact of observation and its influence on group dynamics, paving the way for a deeper understanding of how being watched affects behavior, both at the particle level and in the fabric of human society.

Historical Development of the Observer Concept in Quantum Theory

The notion of the observer in quantum theory has undergone significant transformation since the early 1900s, challenging established ideas about reality and perception. Central to quantum mechanics is the intriguing concept that particles remain in a state of potentiality until observed, a premise introduced by pioneers like Max Planck and Albert Einstein. However, it was the contributions of Niels Bohr and Werner Heisenberg, through the Copenhagen interpretation, that profoundly shaped our understanding. This interpretation suggests that observation collapses a particle's wave function into a definite state, implying that reality is not fixed but rather shaped by the conscious act of observation. This idea continues to spark fascinating debates regarding the nature of existence.

Experiments such as the double-slit test have solidified the observer effect as a fundamental aspect of quantum mechanics. In this experiment, particles like electrons or photons exhibit wave-like properties, forming an interference pattern when not observed. Yet, upon observation, they behave like discrete particles, collapsing the interference pattern. These findings highlight the significant impact of observation, suggesting that merely measuring or witnessing can alter particle behavior. This raises thought-provoking questions about the role of consciousness in shaping reality and whether the universe exists independently of observation or is inherently linked to the observer's presence.

Theoretical interpretations of the observer effect have sparked vibrant discussions within the scientific community, leading to varied perspectives. Hugh Everett III's many-worlds interpretation offers a bold alternative to the Copenhagen interpretation by suggesting that all potential outcomes of a quantum event exist simultaneously in parallel universes, eliminating the need for wave function collapse. Conversely, the pilot-wave theory, developed by Louis de Broglie and later expanded by David Bohm, proposes that particles follow deterministic paths, guided by a hidden wave function unaffected by observation. These interpretations enrich the landscape of quantum mechanics, emphasizing that the observer effect remains an area ripe for exploration and reinterpretation.

The implications of the observer effect extend beyond quantum physics, touching on profound philosophical questions about reality and consciousness. If observation shapes the physical world, then reality itself might be inherently subjective, dependent on the observer's consciousness. This perspective challenges the notion of an objective universe, prompting a reevaluation of philosophical paradigms that have long dominated Western thought. The idea that consciousness plays a central role in the universe invites interdisciplinary dialogue, bridging quantum physics with fields like cognitive science and philosophy to explore the mysterious relationship between mind and matter.

Reflecting on the historical evolution of the observer concept reveals broader implications for our understanding of reality. The observer effect challenges us to reconsider the boundaries between objective and subjective experiences, urging us to question the very fabric of existence. As we delve deeper into the quantum realm, the insights gained from this exploration have the potential to revolutionize not only our understanding of the universe but also the nature of human perception and consciousness. By embracing the complexities and contradictions inherent in quantum mechanics, we open ourselves to a richer, more nuanced appreciation of the world and our place within it.

Key Experiments Demonstrating the Observer Effect in Quantum Mechanics

The interplay between observation and quantum mechanics captivates both scientists and philosophers, forming a fundamental pillar of quantum theory. This intriguing phenomenon proposes that merely observing a quantum system can change its state, challenging traditional ideas of objectivity. A key experiment illustrating this is the double-slit experiment. When particles like electrons pass through a barrier with two slits, they produce an interference pattern on a screen behind, resembling water waves. Yet, if these particles are observed, the pattern vanishes, and they act like classical particles, hitting the screen as if they've traveled through one slit or the other. This experiment prompts profound questions about reality, hinting that particles remain in a superposition of states until observed.

Beyond the double-slit experiment, the quantum Zeno effect further demonstrates how observation impacts quantum states. Named after Zeno, an ancient Greek philosopher, this effect indicates that continuous observation can slow or even stop a quantum system's progression. Consider a radioactive atom about to decay; if it's constantly observed, the decay process can be postponed. This concept reveals the unique nature of quantum systems, where observation seems to freeze particle motion, reminiscent of Zeno's paradox of the arrow that appears motionless when continuously watched.

Despite strong evidence supporting the observer effect, debates about its implications continue. Some, like the Copenhagen interpretation, suggest consciousness collapses a quantum wave function, implying a profound connection between mind and matter. On the other hand, the many-worlds interpretation argues that every possibility unfolds in parallel universes, with observation merely selecting which universe we experience. These varied views underscore the richness of quantum mechanics, encouraging exploration of the limits of knowledge and perception.

Advances in quantum technology offer new ways to explore the observer effect. Quantum computing, for example, uses qubits in superposition, enabling calculations beyond classical limits. As researchers develop more advanced quantum experiments, they delve deeper into the relationship between observation and reality, potentially reshaping our understanding of existence.

This evolving field invites a reevaluation of knowledge and the role of observation in shaping the universe.

These inquiries also prompt consideration of whether similar principles extend beyond the quantum realm. If observation influences particle behavior, what implications does this have for human interactions and social dynamics? Could being observed alter our behavior similarly to quantum systems? These questions encourage reflection on social contexts, inspiring a deeper examination of the connection between observation and action, potentially transforming our understanding of both the quantum world and human society.

Theoretical Interpretations and Debates Surrounding Observers in Quantum Physics

The observer effect in quantum physics presents a perplexing challenge that has ignited extensive debate among scientists and philosophers. At its essence, this phenomenon reveals how simply observing a quantum system can change its state, indicating a significant link between measurement and the nature of reality. This idea contradicts the classical notion of a universe existing independently of observation, suggesting instead that observation is integral in forming the physical world. The Copenhagen interpretation, for example, suggests that particles remain in a state of probability until observed, at which point they assume a definite state. This viewpoint has been instrumental in the development of modern quantum mechanics, yet it remains a topic of lively discussion as scholars delve into the impact of consciousness and measurement on reality's structure.

Numerous pioneering experiments have demonstrated the observer effect, with the double-slit experiment standing out as particularly notable. In this setup, particles such as electrons display wave-like characteristics when not observed, forming an interference pattern. Once observed, they act like particles, and the interference pattern vanishes. This puzzling change underscores the observer's active participation, prompting inquiries into the nature of reality and its potential subjectivity. Such experiments continue to test the limits of traditional physics, inviting fresh perspectives on how observation connects to existence. Recent breakthroughs in quantum research have broadened our understanding, indicating that the observer's role might extend beyond the quantum level, influencing larger systems.

Different interpretations of the observer effect have emerged, each offering distinct insights into the act of observation. The many-worlds interpretation, for instance, posits that all potential outcomes of a quantum event coexist in a multiverse, with observation only selecting which reality we experience. This theory contrasts with the Copenhagen interpretation by suggesting that reality consists of a multitude of possibilities rather than a single outcome. Other perspectives, like the relational interpretation, argue that the effects of observation are relative to the observer, challenging our notions of objectivity. These interpretations highlight the evolving landscape of quantum theory, where emerging ideas continually question established paradigms and prompt a reevaluation of our understanding of the universe.

As quantum theory advances, so does our grasp of its implications for consciousness and reality. Some modern thinkers consider whether the observer effect implies a deeper connection between mind and matter, blurring the lines between physics and philosophy. The potential role of consciousness in shaping reality has gained attention, leading to interdisciplinary studies that merge quantum mechanics with cognitive science. These explorations investigate whether consciousness itself might be a quantum phenomenon, affecting perceptions of reality at a fundamental level. Such discussions encourage us to reconsider longstanding beliefs about existence and the possible interdependent relationship between observer and observed, challenging traditional notions of separateness.

Ongoing debates and research surrounding the observer effect not only enhance our understanding of quantum mechanics but also inspire practical applications in technology and social sciences. In the field of quantum computing, the observer effect is crucial for error correction and information processing, pushing the boundaries of what computers can achieve. In the social domain, drawing parallels between quantum observation and human interactions offers innovative ways to analyze behaviors and decision-making processes. By recognizing the impact of observation, we can develop more nuanced models of social dynamics, improving our ability to predict and influence outcomes in complex systems. This intersection of quantum physics and social science offers fertile ground for future exploration, promising to reshape our understanding of both the microscopic and macroscopic worlds.

Implications of the Observer Effect for Understanding Reality and Consciousness

The observer effect in quantum physics reveals an intriguing relationship between observation and reality, challenging our perception of both the physical world and consciousness. In the quantum realm, measuring a particle can change its state, implying that reality is not fixed but can be shaped by observation. This concept has triggered intense debates about the observer's role, questioning whether consciousness is crucial in forming the universe. These ideas urge us to rethink reality, where observing plays an active role in events rather than merely documenting pre-existing conditions.

Recent quantum research has highlighted the subtle connection between the observer and the observed, offering new insights into how this interaction might reflect human consciousness. The double-slit experiment, central to quantum theory, shows particles behaving like waves until observed, at which point they take on particle-like characteristics. This suggests that observation might trigger potential realities to manifest. Similarly, consciousness's role in perceiving these changes has become a rich area for exploration, with some theorists suggesting it may play a key role in shaping the physical world.

Drawing connections between quantum observation and human consciousness, one can speculate on implications for understanding the mind's influence on personal and collective experiences. Just as quantum particles are affected by observation, human thoughts and perceptions may mold reality in ways we're just beginning to grasp. This viewpoint encourages reevaluating the power of human awareness, proposing that our thoughts and intentions might not only reflect our reality but actively create it. Viewing consciousness as a co-creator of reality opens up possibilities for exploring how mindfulness and intentionality could influence both personal and societal outcomes.

As researchers explore the link between the quantum observer effect and consciousness, emerging theories suggest that consciousness could be a fundamental aspect of the universe. The idea that the mind's awareness can affect reality invites a radical rethink of philosophical and scientific frameworks. This perspective aligns with some interpretations of quantum mechanics proposing a more interconnected universe, where the observer and the observed are inseparably linked. Such ideas challenge traditional views of an objective reality independent of perception, suggesting a more symbiotic relationship between consciousness and the cosmos.

Considering these profound connections, one might wonder about the practical applications of understanding consciousness through a quantum lens. If our awareness truly impacts reality, techniques to harness this power could transform various life aspects. Mindfulness practices, visualization strategies, and conscious intention-setting may help align personal and collective realities with desired outcomes. As the lines between the observer and the observed blur, adopting this quantum perspective could inspire innovative approaches to problem-solving, creativity, and fostering a deeper sense of connection with the world.

How Being Watched Alters Behavior: Quantum Parallels in Human Interaction

Observing the world around us can lead to profound insights, not just in the enigmatic domain of quantum physics but also in our everyday social lives. In particle physics, the act of watching can change how particles behave, hinting at a reality that's altered simply by being noticed. This intriguing phenomenon extends beyond the microscopic; humans, too, adjust their actions when aware of being watched. This shift mirrors the concept of observation impacting behavior, influencing our social exchanges and relationships. This parallel encourages us to explore how the awareness of being seen shapes our thoughts and interactions with others.

As we explore this idea further, the psychological impact of being under surveillance becomes clear. It changes not only how individuals act but also the dynamics within groups. The subtle pressure to conform intensifies in the presence of watchful eyes, affecting how we align with societal norms. In a world increasingly monitored by artificial intelligence, the lines between private and public spaces blur, prompting us to reconsider our behaviors. Each aspect of this intricate interaction reveals more about the delicate balance between observer and observed, offering a deeper understanding of how these dynamics influence our social fabric.

The Observer Effect: From Quantum Particles to Social Surveillance

In the captivating world of particle physics, the concept of the observer's impact illustrates how simply watching can change the state of a system. This intriguing idea, often exemplified by subatomic behaviors, finds fascinating parallels in our social interactions, where being watched significantly alters human conduct. The mere presence of a spectator can influence decisions, actions, and even thoughts, creating a dynamic similar to the microscopic phenomenon. This interchange between observation and conduct plays out across various social and cultural contexts, from subtle shifts in personal behavior under surveillance to more pronounced changes within groups. By exploring this complex landscape, we gain vital insights into how observation affects human relationships and societal frameworks.

In social monitoring contexts, the influence of being watched goes beyond mere perception, acting as a strong catalyst for change. Studies consistently reveal that individuals alter their actions when aware of being observed, evident in settings from workplaces to social media. This heightened awareness often leads to increased self-regulation, aligning behaviors with societal expectations. Much like particles changing states when observed, humans tend to adopt behaviors that conform to perceived norms, often driven by a desire for acceptance or fear of judgment, underscoring the profound psychological effects of observation.

The ramifications of this observation-driven change are particularly noticeable in social conformity. When people sense their actions are under scrutiny, they often adhere more closely to group norms, sometimes at the cost of personal beliefs. This mirrors the particle physics world, where elements display different states depending on the observer's presence. The psychological pressure of being watched can lead to conformity, as individuals place group cohesion above personal expression. This dynamic raises questions about the balance between individual autonomy and collective identity, urging us to consider how much of our behavior is self-directed versus influenced by society's unseen gaze.

In today's digital era, artificial intelligence emerges as a new type of spectator, transforming the landscape of privacy and behavior. AI technologies capable of monitoring and analyzing extensive datasets introduce a fresh dimension to the impact of observation. Unlike human watchers, AI provides an impartial, continuous form of scrutiny, prompting individuals to adjust their actions in anticipation of algorithmic evaluation. This shift significantly affects privacy as the lines between public and private actions blur. The pervasive presence of AI surveillance challenges traditional views of observation, forcing us to redefine what it means to be watched in a digital world.

As we ponder the broader implications of being observed, we are led to ask critical questions about our interactions and the forces shaping them. How do we balance security and accountability with personal freedom and authenticity? What role should AI play in our social structures, and how can we ensure it enhances rather than limits our human experience? These questions encourage reflection on the delicate balance between surveillance and autonomy, inviting us to apply these insights practically. By adopting a particle physics perspective on social dynamics, we can better navigate modern life's complexities, fostering environments where observation enriches rather than diminishes human interaction.

Behavioral Shifts: The Psychological Impact of Being Monitored

In the field of quantum physics, the observer effect suggests that simply observing a particle can alter its state. This fascinating concept mirrors human behavior, where the knowledge of being observed can significantly shape an individual's actions and decisions. When people are aware of being watched, they often modify their behavior, consciously or unconsciously, to meet perceived expectations. This phenomenon is supported by extensive psychological research, which indicates that surveillance, whether physical or digital, can increase self-awareness and conformity to social norms. The complex interplay between observer and observed in both quantum mechanics and human interactions highlights the significant influence perception has on reality.

The psychological effects of being monitored extend beyond basic behavioral changes. When individuals sense they are being watched, they often feel a heightened sense of accountability, potentially leading to improved performance and rule-following. This is evident in workplaces with surveillance cameras or classrooms with exam invigilators. While this oversight can enhance efficiency and productivity as people strive to put forth their best selves, the persistent nature of being observed can also cause stress and anxiety, particularly when the observer's intentions are unclear. The constant awareness of being monitored can create a high-pressure environment, where maintaining authenticity becomes challenging.

As technology advances, the nature of observation has evolved. Artificial Intelligence now acts as a modern-day sentinel, capable of monitoring on a vast scale. This digital surveillance has redefined privacy and personal space, challenging individuals to navigate an ever-watchful world. AI systems can analyze behavioral patterns, predict future actions, and even influence decision-making. While AI can enhance trust and accountability, it also raises ethical concerns regarding autonomy and freedom. Balancing AI-powered observation's benefits with personal privacy needs is a critical challenge, prompting society to reconsider what it means to be free in a connected world.

Exploring observation's psychological impact reveals that a watchful presence can drive social conformity. People often mimic the behaviors and attitudes they believe are expected by observers. This is visible in social settings where individuals adopt group norms to fit in, subconsciously aligning their values and actions with the majority. An observer's presence can intensify these tendencies, as the desire for social acceptance becomes more pronounced. This dynamic can lead to homogenized behavior, stifling unique perspectives and creativity in favor of group cohesion. The challenge lies in encouraging environments where diverse thought is valued, even under the gaze of observation.

To navigate the complexities of being observed, individuals and societies need a nuanced understanding of the observer effect. A practical approach involves fostering environments where transparency and trust are prioritized, balancing accountability with autonomy. Open discussions about surveillance's impact can empower individuals to make informed choices about their behavior. Mindfulness practices can help maintain authenticity under observation, enabling actions aligned with true values rather than external pressures. Recognizing observation's dual nature as a catalyst for positive change and a potential stress source allows society to harness its potential to create environments that are both productive and psychologically safe.

Social Conformity and the Presence of a Watchful Eye

Social conformity is a widespread phenomenon evident across various cultures and communities, drawing intriguing parallels to the quantum principle of observation. Just as particles exhibit different behaviors when observed, individuals often alter their actions in the presence of others. This adaptation, motivated by the desire for acceptance and the fear of judgment, highlights the delicate balance between personal individuality and societal norms. When people know they are being watched, whether by peers or authority figures, there is a subconscious pull towards conforming to expected behaviors. This shift underscores the significant influence that mere observation can have on human conduct, echoing the quantum observer effect.

The psychology of social conformity is deeply embedded in our evolutionary past. As inherently social creatures, humans have developed mechanisms to foster group cohesion and survival. In interconnected societies, individuals navigate complex social landscapes where conformity often ensures inclusion and protection. This adaptation is evident in various settings, from workplace norms to cultural rituals. The presence of an observer encourages people to align their actions with the group, promoting unity but sometimes stifling creativity and personal expression. These dynamics show how observation can alter not just actions but potentially the very essence of human identity within a collective.

In today's digital age, the omnipresence of surveillance has magnified the observer effect. Knowing that actions may be monitored online or in smart environments has ushered in a new era of behavior modification. People carefully craft their online personas, often presenting an idealized version of themselves. This digital observer effect can increase the pressure to conform to societal ideals perpetuated through social media and other digital platforms. However, it also offers opportunities for self-reflection and growth, as individuals become more aware of the impressions they create and the norms they choose to embrace or challenge.

Within this digital realm, AI emerges as a transformative force, reshaping the role of the observer. Unlike human watchers, AI systems can analyze patterns and behaviors on an unprecedented scale, providing insights into social conformity previously unattainable. By examining extensive datasets, AI can identify subtle trends in how observation influences behavior across different contexts and cultures. This ability not only enhances our understanding of social dynamics but also raises ethical questions about privacy and the extent to which observation should dictate behavior. As AI continues to evolve, it sparks a reevaluation of the balance between the benefits of observation and the preservation of individual autonomy.

To navigate the complexities of social conformity in an age of heightened observation, individuals can adopt strategies that promote authentic self-expression while remaining aware of societal influences. Cultivating self-awareness and critical thinking allows for conscious choices about when to conform and when to assert individuality. Engaging with diverse perspectives and communities can reduce the pressure to conform, as exposure to varied ideas fosters an appreciation for uniqueness and innovation. Ultimately, understanding the parallels in social behavior provides people with the tools to engage thoughtfully with the world around them, fostering a society that values both unity and diversity.

AI as the New Observer: Redefining Privacy and Behavior

In today's world, artificial intelligence is becoming an ever-present observer, reshaping privacy concepts and impacting human behavior. As algorithms grow more advanced, they monitor and analyze actions, reminiscent of the observer effect in quantum physics, where observing alters a particle's state. This societal parallel sees people changing their actions when aware of being watched, whether by public cameras or online trackers. AI's role as an observer has deep implications, leading to behavioral changes often subconsciously as people adapt to constant surveillance.

The psychological impacts of AI observation are complex. It encourages self-control and adherence to social norms, as people become more aware of their actions under scrutiny, which can be advantageous for public safety by deterring crime. Conversely, the omnipresence of AI surveillance can create discomfort or self-censorship, limiting genuine expression and creativity due to fear of judgment. This creates a tension where the benefits of safety and order may hinder individuality and innovation.

AI's observational role goes beyond surveillance, influencing decision-making processes. By analyzing extensive data, AI forecasts trends and behaviors, subtly guiding choices through targeted ads and personalized content. While this can enhance user experience with tailored recommendations, it also raises ethical concerns about autonomy, as people may unknowingly follow AI-suggested paths, narrowing their perspectives. This balance between guidance and control poses a modern dilemma, challenging society to weigh technological advantages against ethical considerations.

As AI advances, its ability to redefine privacy and behavior will grow. Emerging technologies like facial recognition and emotion detection promise deeper insights into human behavior, leading to more nuanced understanding and interaction models. These advancements hold exciting potential for sectors like healthcare and education, where AI's observational skills can offer personalized interventions and learning experiences. However, these developments demand a critical examination of privacy rights and acceptable surveillance boundaries, urging policymakers and technologists to create frameworks that protect individual freedoms while fostering innovation.

To navigate this complex landscape, individuals can proactively manage their digital footprints and safeguard privacy. Being discerning about sharing personal information, using privacy tools, and staying informed about AI's capabilities empower people to interact with technology on their terms. Additionally, fostering digital literacy and critical thinking can help society adapt to AI as an observer, encouraging a culture where individuals stay informed and proactive in shaping their tech interactions. By adopting a balanced approach considering both opportunities and challenges of AI surveillance, society can harness its potential while preserving essential human values.

The Social Observer Effect: How Group Dynamics Shift Based on Who's Watching

Imagine stepping into a vibrant room buzzing with conversation, where interactions flow effortlessly. Suddenly, the dynamic shifts as a person of significance enters, causing laughter to quiet and postures to change, with everyone acutely aware of their presence. This scenario mirrors the fascinating parallel to the quantum realm, where particles exhibit different behaviors when observed. The mere awareness of being watched can influence people's choices and actions, often in ways that aren't immediately obvious. By delving into these subtle shifts, we uncover how social frameworks adapt under observation, revealing the intricate interplay between perception and reality.

Delving deeper, the influence of social status emerges as a vital factor, highlighting how hierarchy can dictate group behavior. Visibility introduces another dimension, prompting individuals to either conform or resist when under public scrutiny, adding complexity to decision-making. The perceived weight of surveillance can either fortify group unity through shared awareness or cause it to splinter under pressure. Additionally, the influence of external factors offers new insights into group dynamics, demonstrating the unpredictable manner in which human interactions can resemble quantum phenomena. Through this lens, we explore the social observer effect, offering a glimpse into the unseen forces shaping our collective experiences.

The Role of Social Status in Modulating Group Behavior

Social status exerts a substantial influence on group dynamics, acting as a powerful yet often unseen force that shapes interactions and decisions within a collective. Recent findings in social psychology highlight that individuals with elevated status frequently set behavioral standards for others to emulate. This dynamic is reminiscent of the observer effect in quantum physics, where the mere presence of an observer alters the behavior of particles. Similarly, in group settings, the presence of high-status individuals can lead members to modify their actions and attitudes, often in an unconscious pursuit of validation and acceptance. This tendency fosters conformity as individuals align their behaviors with influential figures.

In environments where social status is a pivotal element, the presence of a prominent person can significantly alter group dynamics. For instance, a CEO or respected leader's participation in workplace meetings or social gatherings often shifts the group's focus and impacts decision-making processes. These individuals wield considerable influence, with their opinions often weighing more heavily than others. This power imbalance can result in "groupthink," where the desire for consensus overrides rational decision-making. Recognizing the influence of status on interactions can aid organizations in creating inclusive environments where diverse perspectives are valued, reducing the risk of uniformity in thought and action.

The influence of social status on group behavior extends to digital realms, where visibility and perceived esteem are crucial in shaping online interactions. On social media platforms, users with large followings frequently become trendsetters, with their endorsements or criticisms influencing public opinion and behavior. These digital influencers, akin to their real-world counterparts, create significant impacts within social networks, highlighting how status transcends physical boundaries to affect virtual communities. The algorithmic promotion of high-status individuals' content further amplifies their influence, creating a cycle that reinforces their authority. Understanding these patterns can guide the design of digital platforms to promote diverse voices and equitable representation.

The psychological roots of status-driven behavior offer a rich area for exploration. Neuroscientific research suggests that humans are inherently responsive to hierarchical cues, with studies indicating that social status affects brain activity related to reward processing and empathy. This intrinsic sensitivity points to an evolutionary basis for status-driven group dynamics, where aligning with higher-status individuals may have historically provided survival advantages. By understanding these ingrained tendencies, we can devise strategies to counteract the adverse effects of status hierarchies, promoting environments that encourage collaboration and innovation over deference and division.

To challenge traditional power structures, individuals and groups can implement strategies to democratize influence and encourage diverse contributions. Initiatives such as rotating leadership roles, fostering open dialogue, and valuing merit-based contributions can disrupt conventional hierarchies. Encouraging critical thinking and questioning established norms empowers individuals to act authentically rather than merely conforming to the expectations set by high-status figures. By addressing the role of social status in group behavior, we can unlock the potential for more dynamic and equitable social structures, paving the way for innovation and progress.

Visibility and Conformity: How Public Scrutiny Alters Decisions

The impact of visibility significantly shapes both individual and group decision-making, drawing intriguing connections between quantum mechanics and social interactions. When people are aware of being observed, their behavior often changes, similar to how particles behave in a quantum setting when watched. This phenomenon, akin to the quantum observer effect, implies that awareness of being scrutinized can lead to increased conformity, altering group dynamics. Recent research highlights how public scrutiny can push individuals to align with perceived social norms, reducing deviations and creating a sense of uniformity. This behavioral shift, driven by the desire for social acceptance, underscores the powerful role of visibility in decision-making processes.

As awareness of observation grows, the tendency to conform intensifies, reflecting a subtle psychological interplay. The presence of an observer, whether real or imagined, often triggers heightened self-monitoring, where individuals carefully assess their actions against expected standards. This self-awareness appears in various situations, from workplace settings to social gatherings, where pressure to follow group norms rises. Social psychologists have studied how this awareness can be a double-edged sword: it can foster harmonious group dynamics but may also suppress creativity and personal expression, limiting the diversity of ideas and actions within the group.

Delving into this social observer effect reveals its potential impact on decision-making in high-stakes environments. In organizational settings, for example, employees might lean toward decisions that reflect institutional norms when under direct observation, potentially stifling innovation. On the other hand, when individuals feel protected from scrutiny, they may feel more confident in expressing dissenting opinions or proposing new ideas. This duality emphasizes the need for balanced environments where visibility is managed thoughtfully to encourage both conformity for cohesion and freedom for innovation. Leaders who grasp this dynamic can create spaces that leverage the benefits of observation without curbing individual initiative.

These dynamics extend beyond personal interactions to broader societal phenomena. The growth of digital platforms has heightened public visibility, changing how groups form and interact. In the realm of social media, users often adjust their behavior based on perceived audience expectations, crafting personas that might differ from their authentic selves. This digital visibility influences not only personal decisions but also transforms public discourse, as individuals navigate the tension between maintaining social credibility and expressing genuine viewpoints. As technology continues to evolve, understanding how visibility impacts decision-making will be crucial for cultivating healthy online communities that balance authenticity with social harmony.

By examining the role of visibility and conformity in shaping decisions, we gain valuable insights into the mechanics of social cohesion and change. Creating environments where individuals feel seen yet free to express diverse ideas can lead to more vibrant and resilient societies. Consider the potential of designing spaces where observation is a tool for growth rather than a constraint, prompting reflection on how visibility can inspire positive change. As we navigate an increasingly interconnected world, embracing the nuanced effects of public scrutiny can empower individuals and groups to make decisions that are both socially conscious and authentically aligned with their values.

The Psychological Impact of Perceived Surveillance on Group Cohesion

The perception of being watched plays a crucial role in shaping how groups function and maintain cohesion. When individuals feel they are under surveillance, they often become more self-aware, resulting in behavior that aligns with group norms rather than personal expression. This behavioral shift echoes the observer effect in quantum physics, where the act of observation affects the phenomenon being observed. As a result, people might adjust their actions to meet the expectations of those they believe are watching, influencing group decision-making and harmony. Recent social psychology research indicates that perceived observation often leads to behavior that matches observer expectations, fostering group unity at the expense of diversity. This understanding urges a reevaluation of surveillance—both actual and perceived—and its dual capacity to either strengthen or weaken group bonds.

In settings like workplaces or educational institutions, where surveillance is commonplace, its impact on group cohesion can vary. On one hand, the knowledge of being watched can instill a sense of accountability, prompting individuals to align with collective goals and standards. On the other hand, excessive scrutiny can suppress creativity and inhibit risk-taking, as people focus on compliance over innovation. Achieving a balance between these outcomes depends on how the intent behind the surveillance is perceived. When viewed as supportive, it can boost group efficacy; when seen as controlling, it can erode trust and cooperation. Organizations must thoughtfully design surveillance policies to enhance positive group dynamics while minimizing negative consequences.

Investigating the psychological mechanisms of perceived surveillance reveals complex layers influencing group cohesion. The "panopticon," a concept introduced by philosopher Jeremy Bentham, illustrates how the mere possibility of observation can regulate behavior. In today's digital age, surveillance tools intensify this effect, as people engage on social media and digital platforms with the awareness that their actions might be monitored. This heightened visibility encourages self-regulation but also raises concerns about authenticity and the potential loss of genuine connections. The challenge is to create environments where people feel safe to express themselves authentically despite the constant presence of potential observers.

Consider a workplace where employees believe management closely monitors their interactions. This perception can foster conformity and a superficial unity as employees strive to meet expected norms. However, this enforced uniformity might mask underlying issues such as unresolved conflicts or innovative ideas that deviate from the norm. Encouraging open dialogue and feedback mechanisms can help counter these effects, allowing groups to benefit from perceived surveillance while preserving diverse voices and perspectives. By cultivating an environment of trust and transparency, organizations can leverage the positive aspects of surveillance to bolster group cohesion without sacrificing individuality.

The interaction between perceived surveillance and group cohesion provides valuable insights for those navigating modern social environments. Recognizing the dual nature of surveillance and its potential to both unite and divide, stakeholders can adopt strategies that foster a culture of respect and collaboration. This involves understanding the psychological impact of surveillance and actively engaging with diverse perspectives to build resilient, cohesive groups. As society continues to face the challenges and opportunities of an increasingly observed world, these insights are essential for creating environments where individuals and groups can thrive together.

Emergent Group Dynamics in the Presence of External Influences

In the sphere of group dynamics, external factors can significantly influence new behavioral patterns. An intriguing dimension of this is how groups, when under observation, may unconsciously adjust their collective behavior to meet perceived expectations. This phenomenon can be compared to particles in quantum physics that change states when observed, creating a fascinating link between physics and social science. A recent study utilizing AI-driven analysis examined workplace teams under varying observational conditions. The findings showed that the mere suggestion of being monitored altered communication styles, decision-making, and risk-taking within the groups. These insights highlight the profound impact external scrutiny can have on group dynamics, prompting a reassessment of how observation influences social behavior fundamentally.

Extending this concept, external influences are not limited to physical observation but also include digital surveillance and social media presence. In our interconnected world, the knowledge that digital activities are constantly watched can cause shifts in group dynamics. Social platforms serve as modern observers, impacting group interactions and decisions. For example, a team working on a project might change their strategies if aware that their progress is shared online. This awareness can lead to positive outcomes, such as heightened accountability and collaboration, but also negatives like conformity and groupthink. By understanding these dynamics, organizations can leverage the benefits of surveillance while addressing its downsides, fostering environments where innovation and creativity flourish.

The psychological basis of perceived surveillance plays a crucial role. When individuals feel observed, they naturally conform to socially accepted norms due to the desire for approval and fear of negative judgment. This can suppress creativity and discourage dissenting opinions, leading to homogenized thinking. To counteract this, leaders can cultivate an atmosphere of trust and openness, where diverse opinions are valued, and judgment is minimized. By fostering such an environment, the negative effects of perceived surveillance can be mitigated, allowing for a more authentic and innovative group dynamic.

The relationship between external observation and group behavior offers valuable insights into collective intelligence. Groups observed to perform more effectively often show increased coherence and synchronization in their interactions, akin to quantum coherence, where particles exist in harmony. Practically, groups functioning under observation with a shared understanding and common goals tend to outperform others. By strategically applying this understanding, organizations can create conditions that enhance group coherence, leading to better problem-solving and agility in facing challenges.

To apply these insights, individuals and organizations can take steps to optimize group dynamics when influenced by external factors. Transparency about observation practices, both digital and physical, helps reduce anxiety and demystifies the process. Establishing clear objectives and fostering open communication can empower groups to adapt positively to observational pressures. Recognizing the potential for external influences to both hinder and enhance group dynamics, astute leaders can navigate these complexities to build resilient, adaptable teams capable of thriving in an ever-evolving social landscape.

In the intricate world of particle physics, the impact of observation underscores a fascinating reality: watching a phenomenon can alter its state. This chapter has shed light on how this principle resonates within human social interactions. The awareness of an audience can significantly influence actions, decisions, and social norms, whether in one-on-one encounters or complex group settings. As we explore these parallels, the influence of observation challenges us to rethink how our perceptions and focus shape the social environment. This understanding deepens our awareness of our dual roles as observers and participants in society, encouraging readers to reflect on their influence and presence in social situations. As our journey through this subatomic perspective continues, the insights gained invite a thoughtful engagement with the unseen forces at play, urging us to interact with the world more mindfully.In the fascinating world of subatomic science, the watcher effect holds a revealing insight: simply observing can transform the state of what is being viewed. This chapter has shed light on how this concept echoes within human interactions, where the awareness of being observed can significantly influence and reshape behaviors.

Chapter Five

Quantum Tunneling And Breaking Social Barriers

Imagine a world where the rigid boundaries that typically govern our social interactions begin to fade, similar to how subatomic particles can unexpectedly traverse barriers in quantum physics. This is the essence of quantum tunneling—a phenomenon that challenges traditional logic and encourages us to reconsider the steadfast limits we face in daily life. As we venture into this concept, we discover that the behavior of particles at the microscopic level offers a powerful analogy for understanding how individuals can break through societal conventions and expectations that often hold us back.

Consider an artist who, defying the odds, emerges from a strict cultural background to inspire a transformative artistic movement, or a trailblazing leader who dismantles the glass ceiling in a male-centric industry. These stories are not merely personal victories; they embody a fundamental principle, much like quantum tunneling allows particles to materialize beyond barriers without the requisite energy for conventional passage. This chapter draws connections between the physical and social realms, illustrating how the principles governing the universe's smallest components can illuminate pathways for individuals and collectives to challenge and reshape the status quo.

As we delve further, we acknowledge the transformative role of artificial intelligence in this landscape. By sifting through vast datasets and uncovering patterns invisible to the human eye, AI acts as a modern-day guide, forecasting when and how social breakthroughs might occur. It provides insights into behavioral shifts that can catalyze profound societal transformations, akin to a compass navigating us through unexplored territory. In this chapter, we will explore these concepts and their implications for our understanding of human interactions, urging readers to observe the world through a lens that is both scientific and deeply human.

Quantum Tunneling: Overcoming Barriers in Physics

What allows particles to seemingly defy logic and traverse barriers that should be impassable? This extraordinary occurrence in the quantum world, known as quantum tunneling, challenges our conventional understanding of obstacles. Picture a scenario where an electron, rather than being halted by a barrier, effortlessly appears on the other side, as if the blockade never existed. This phenomenon not only disrupts traditional physics but also prompts us to rethink how we view limitations in our social lives. Just as particles maneuver through obstacles, individuals can navigate and transcend societal conventions, redefining what we consider achievable. The parallels between these realms urge us to contemplate how unseen forces of determination and creativity can propel people beyond society's invisible constraints.

In this chapter, we delve into the captivating concept of quantum tunneling and draw meaningful connections to human behavior and societal transformation. As we explore the subatomic world, we'll discover analogies that reveal how individuals break through established norms, much like particles overcoming physical barriers. These insights lay the groundwork for understanding how AI might offer new ways to model and anticipate such social breakthroughs, hinting at potential shifts in behavior that could transform communities and cultures. Each section will unravel the layers of this intricate interplay, illustrating how quantum principles can inspire social innovation and collective advancement. As we embark on this exploration, the interaction between particles and people invites us to imagine new possibilities for navigating and reshaping our social landscapes.

The Principle of Quantum Tunneling in Subatomic Particle Behavior

Quantum tunneling presents a fascinating phenomenon where particles defy classical expectations by moving through energy barriers that seem impossible to cross. This process highlights the unique aspects of quantum mechanics, where particles like electrons exhibit wave-like behavior, existing in a state of potentiality. This probabilistic nature allows them to appear on the other side of a barrier without climbing over it in the traditional sense. Such behavior challenges our classical understanding and offers profound insights into the fundamental workings of matter.

Recent progress in quantum mechanics has expanded our understanding of tunneling, with state-of-the-art experiments revealing its involvement in various natural processes. One notable example is nuclear fusion in stars, where protons overcome repulsive forces and fuse, releasing immense energy—a process partially facilitated by tunneling. Moreover, this phenomenon is essential to technologies like the scanning tunneling microscope, enabling scientists to view surfaces at the atomic level. These instances emphasize the widespread relevance of tunneling, illustrating its importance beyond theoretical confines and its role in advancing human knowledge.

Drawing parallels between quantum tunneling and human behavior opens interesting avenues for interpreting social dynamics. Just as particles can unexpectedly navigate barriers, individuals and groups often transcend societal norms and constraints in surprising ways. This analogy offers a fresh perspective on social innovation, where entrenched customs are re-evaluated and reshaped. The unpredictable nature of such breakthroughs mirrors the quantum realm, where certainty often eludes us, and potentialities are vast. By embracing this viewpoint, we gain a richer understanding of the complexities and possibilities inherent in human interactions, fostering a more nuanced comprehension of societal evolution.

Artificial intelligence significantly contributes to modeling these quantum-inspired dynamics, providing new tools for analyzing and predicting social change. Through machine learning algorithms, researchers can identify patterns in data that hint at where and how social breakthroughs might occur. This predictive ability holds promise for anticipating shifts in cultural norms or the rise of transformative social movements, empowering policymakers and leaders to respond proactively. The fusion of AI with quantum principles enriches our toolkit for addressing complex social phenomena, promoting a more adaptable and responsive society.

Understanding quantum tunneling's implications for social innovation encourages us to rethink how we approach obstacles in our personal and collective lives. By drawing inspiration from the quantum realm, where boundaries are fluid and possibilities endless, we can cultivate a mindset that values openness and flexibility. This perspective invites us to question the rigidity of social structures and embrace novel solutions that might initially seem improbable. In doing so, we not only expand our understanding of the world but also welcome the potential for transformative change in our lives and communities.

Analogies of Quantum Tunneling in Breaking Social Norms

In the mysterious domain of quantum physics, particles defy traditional laws by seemingly traversing energy barriers, a puzzling process called quantum tunneling. This fascinating phenomenon mirrors the human ability to challenge deep-rooted social norms, highlighting the interplay between personal agency and societal limits. Social tunneling emerges when individuals or groups question and dismantle existing conventions, paving the way for innovation and change. This rebellion, reminiscent of particles overcoming obstacles, underscores the hidden potential within communities to transform and redefine collective boundaries. Social tunneling serves as a metaphor for understanding how groundbreaking ideas infiltrate and reshape rigid social structures from within.

Consider the historical changes driven by civil rights movements, which illustrate social tunneling in practice. These movements, fueled by individuals who bravely confronted discriminatory norms, demonstrate how continued efforts can slowly wear down significant societal obstacles. Much like quantum tunneling, social change is probabilistic, often requiring multiple attempts before success is achieved. This aspect of social tunneling highlights the value of perseverance and resilience when facing systemic resistance. By exploring these parallels, we can gain insights into the mechanisms that facilitate substantial societal progress, inspiring innovative strategies to tackle modern challenges.

Recent interdisciplinary research has broadened our understanding of these parallels, integrating insights from sociology, psychology, and quantum physics to create sophisticated models of social tunneling. These models investigate the circumstances under which individuals are most likely to challenge norms, focusing on aspects such as group dynamics, cultural context, and personal traits. For example, studies indicate that environments promoting diverse viewpoints and encouraging dissent increase the likelihood of social tunneling, much like certain conditions enhance quantum tunneling probability. This interaction between individual choice and environmental factors offers fertile ground for future research, inviting new theoretical and practical explorations.

Artificial intelligence is crucial in this exploration, providing advanced tools for analyzing and predicting social tunneling phenomena. By processing large datasets, AI can uncover patterns and trends beyond human perception, revealing insights into the subtle forces driving societal shifts. Machine learning algorithms, for instance, can simulate various scenarios to predict how changes in policy or public sentiment could influence social norms. This predictive power not only deepens our understanding of social dynamics but also equips policymakers and leaders with data-driven strategies to foster positive change, emphasizing AI's role as a catalyst for social innovation.

Readers are encouraged to reflect on how they might apply these insights to their own lives and communities, considering the barriers they wish to overcome and the conventions they aim to challenge. By adopting a quantum perspective, individuals can transform their relationship with societal constraints, viewing them not as insurmountable hurdles but as opportunities for growth and transformation. This mindset invites proactive engagement, motivating individuals to become agents of change within their spheres of influence. Such reflection and action can spark social tunneling in various contexts, from personal relationships to global movements, underscoring the transformative potential of quantum-inspired thinking.

AI's Role in Modeling Quantum Tunneling and Human Interaction Barriers

Quantum tunneling, a fundamental concept in quantum mechanics, reveals a fascinating phenomenon where subatomic particles cross energy barriers that classical physics labels as insurmountable. This idea finds an intriguing parallel in human social interactions, where individuals face and sometimes overcome daunting societal norms and expectations. In this context, artificial intelligence becomes a powerful ally in modeling these intricate dynamics. By simulating scenarios where individuals metaphorically navigate through social constraints, AI sheds light on how obstacles are overcome, offering a deeper understanding of both individual and collective behavioral shifts.

Advanced machine learning algorithms, particularly those specializing in pattern recognition and predictive analytics, exhibit a unique ability to discern and interpret the subtle, often overlooked indicators of social navigation. These algorithms analyze vast datasets from social media, behavioral studies, and historical societal shifts, pinpointing recurring themes and patterns that precede significant social transformations. For instance, AI has been employed to detect emerging cultural trends and shifts in public sentiment, often before they surface in mainstream discourse. By recognizing these early signals, researchers can anticipate when and how social boundaries might be challenged and potentially redefined.

A compelling example is found in the realm of social media, where AI algorithms can monitor the spread of unconventional ideas or behaviors that initially exist on the fringes but gradually gain acceptance. This is akin to observing particles in a quantum state transitioning from one side of a barrier to another. As AI tracks these shifts, it can offer valuable predictions about which movements or ideas are likely to transcend societal resistance and gain widespread acceptance. This ability not only aids in anticipating cultural evolutions but also supports proactive policy-making and strategic planning across various sectors, from education to corporate governance.

The use of AI in modeling quantum-inspired transitions within human interactions goes beyond mere prediction. It provides a framework for fostering innovation and social change. By identifying and understanding the mechanisms through which social norms are challenged, AI can assist in designing interventions or initiatives that encourage positive transformation. For instance, organizations aiming to promote diversity and inclusion can leverage these insights to create environments where unconventional ideas are not only embraced but flourish, effectively lowering the hurdles that traditionally impede progress.

As AI continues to evolve, its role in modeling quantum transitions and human interaction barriers will likely expand, offering more sophisticated tools for understanding and influencing social dynamics. This progress invites reflection on how these insights can be practically applied. Imagine scenarios where AI-driven analyses prompt us to question entrenched societal norms and explore alternative pathways for social evolution. By fostering a mindset open to transcending barriers, individuals and communities can cultivate a more adaptive and resilient social fabric. Ultimately, the convergence of AI and quantum-inspired thinking holds promise for redefining not just our understanding but also our approach to the challenges and opportunities within the complex landscape of human interactions.

Implications of Quantum Tunneling for Social Innovation and Change

Quantum tunneling, where particles pass through obstacles deemed insurmountable by classical physics, serves as a compelling metaphor for societal innovation and transformation. This quantum phenomenon parallels the breaking of societal norms, envisioning a world where conventional barriers dissolve, paving the way for groundbreaking advancements. In society, individuals and groups often confront entrenched obstacles—cultural, institutional, or psychological—that impede progress. Just as quantum particles defy expectations by navigating barriers, societies can transcend long-standing norms to foster innovation and dynamic change. This perspective invites a reevaluation of perceived possibilities, encouraging individuals to question which barriers are genuine and which are merely outdated constructs.

Recent studies in quantum physics and social sciences provide fascinating insights into how societies might emulate tunneling behaviors to drive change. Behavioral science suggests that when individuals perceive a path through seemingly impermeable barriers, they are more likely to challenge and transcend societal conventions. This mirrors quantum tunneling, where the mere possibility of passage alters outcomes. The role of AI in this context is particularly valuable, as it can model and predict these tunneling-like behaviors in social systems, offering glimpses into potential futures where societal innovation is not just possible but inevitable. AI can identify patterns and variables that, when adjusted, enable new pathways for social evolution, akin to the unpredictable trajectories of subatomic particles.

These insights offer practical frameworks for encouraging social progress. By recognizing and understanding the conditions under which social tunneling occurs, policymakers, educators, and leaders can create environments that nurture innovation. For instance, fostering open-mindedness and flexibility within organizational cultures can facilitate the breaking of conventional norms, much like lowering energy barriers in quantum systems. This approach aligns with modern management strategies that emphasize agility and adaptability, drawing inspiration from the principles of quantum mechanics to inspire real-world change.

Countering the mainstream narrative that societal change is slow and arduous, the quantum tunneling analogy offers a fresh perspective on rapid transformation. It challenges the notion that change must be gradual, suggesting instead that by aligning conditions correctly—much like tuning parameters in quantum experiments—societal evolution can be both swift and profound. This concept encourages readers to consider their potential to act as catalysts within their communities, questioning which barriers they might penetrate to instigate meaningful change.

To engage with this paradigm, readers might reflect on societal norms they perceive as barriers and consider how they can apply quantum tunneling principles in their lives. What small, seemingly inconsequential actions could lead to breakthroughs? How can AI tools be harnessed to identify and challenge these barriers? By embracing a quantum mindset, individuals and groups are empowered to redefine the dynamics of their social interactions and contribute to a more innovative and equitable society. Through the lens of quantum tunneling, the path to social change is not only visible but within reach, inviting us all to become pioneers in our personal and collective journeys.

Social Tunneling: How Individuals Break Through Social Norms

In recent times, there's been a growing fascination with the mysterious phenomenon of quantum tunneling, where particles seemingly overcome insurmountable barriers. This intriguing concept from the subatomic world mirrors societal dynamics, where individuals challenge long-standing norms and cultural expectations. Much like particles navigating through physical obstacles, people maneuver through the rigid frameworks of social conventions, occasionally defying conventional reasoning. The social landscape is filled with invisible hurdles—norms dictating behavior, boundaries confining identity, and expectations shaping choices. History is rich with examples of those who have transcended these confines, unexpectedly reshaping societal structures.

Examining this compelling analogy, the idea of social navigation emerges as a powerful metaphor for instigating change, driven by individuals bold enough to question the status quo. Our exploration begins by investigating how quantum principles can shed light on the process of overcoming social stagnation. Cognitive strategies, resembling paths through a labyrinth, reveal the mental frameworks empowering individuals to confront and surpass these challenges. Interconnected identities, akin to entangled particles, influence social interactions, sparking ripples that propel transformation. The rise of AI-driven models offers a fresh perspective to predict and comprehend these behavioral shifts, providing an unparalleled glimpse into the mechanisms of social evolution. As we delve into these themes, the intricate interplay between quantum theories and social change unfolds, uncovering the profound ways they inform and transform each other.

The Role of Quantum Tunneling in Overcoming Social Inertia

Quantum tunneling is a fascinating phenomenon where particles, defying classical physics, permeate energy barriers that seem insurmountable. This concept serves as a compelling metaphor for how individuals can navigate the inertia of societal norms. Much like particles overcoming obstacles in unexpected ways, people often challenge deeply rooted societal conventions. This process, which we can term social tunneling, involves individuals or groups pushing through established norms, even against formidable odds, to instigate change. It highlights the power of human agency and creativity in transcending societal constraints, akin to how quantum particles demonstrate the probabilistic nature of reality.

In social settings, overcoming inertia mirrors the energy required for quantum tunneling. This energy translates into the motivation, resilience, and strategic thinking essential for challenging and reshaping norms. Cognitive strategies are pivotal as individuals maneuver through complex social landscapes, often encountering resistance or backlash. By employing techniques such as reframing narratives, leveraging social networks, and building alliances, people can effectively challenge the status quo. This strategic navigation parallels quantum tunneling, where particles find unlikely pathways through barriers.

Entangled identities introduce additional complexity to the social tunneling process. In quantum physics, entangled particles remain connected regardless of distance, influencing each other's states. Similarly, human identities are often intertwined with those of communities and cultures, affecting how individuals perceive and enact social change. When people recognize their interconnectedness and draw on the strength of collective identities, they can initiate substantial social transformations. This entanglement can amplify the impact of individual actions, creating ripples of change that extend far beyond the initial act of defiance.

Emerging AI-driven models are becoming invaluable tools for understanding and predicting social tunneling and behavioral shifts. By analyzing vast amounts of social data, these models can identify patterns and trends that signal potential for change. AI's ability to simulate various scenarios provides insights into how certain actions might influence social systems. This predictive power can assist individuals and organizations in strategically planning their interventions, enhancing the likelihood of successful norm-breaking initiatives. As AI continues to advance, its role in illuminating the dynamics of social tunneling will undoubtedly grow, offering new avenues for fostering societal progress.

The interplay between quantum tunneling and social change invites individuals to contemplate their role in reshaping societal norms. Reflecting on the courage and innovation required to challenge conventions prompts questions about the barriers one might wish to overcome in their own life or community. What norms persist unchallenged in your environment, and what strategies could you employ to navigate these obstacles? Embracing the metaphor of quantum tunneling empowers individuals to envision themselves as agents of change, capable of transcending limitations and contributing to a more dynamic and equitable society.

Navigating the Barrier: Cognitive Strategies for Challenging Norms

Understanding and navigating the complex environment of societal norms demands inventive cognitive approaches reminiscent of how quantum particles overcome barriers. Just as particles in quantum physics can traverse energy obstacles deemed insurmountable, individuals can creatively, resiliently, and strategically challenge cultural standards. Cognitive flexibility is crucial here, allowing individuals to identify alternative paths and opportunities where others may see limitations. This mental agility enables the exploration of new social behaviors, gradually weakening inflexible norms. Recent psychological research indicates that engaging in exercises that promote divergent thinking can significantly boost one's capacity to navigate and question entrenched social expectations.

In this intricate process of societal transformation, self-awareness and introspection are essential. Individuals need to examine their beliefs and biases, questioning the origins of their perceptions. By recognizing the societal constructs influencing their worldview, they can embark on an introspective journey similar to a particle oscillating between states, moving between conforming to norms and seeking independence. Cognitive-behavioral techniques like reframing and mindfulness have proven effective in helping individuals develop the mental strength to question and surpass social boundaries. These methods enable people to reinterpret social cues and redefine their interactions, fostering a mindset that is both critical and open to change.

The complexity of entangled identities further complicates the dynamics of societal transformation. Much like quantum entanglement defies classical explanations, individual identities are often intertwined with others, affecting their ability to challenge norms. Social identities, influenced by factors such as culture, ethnicity, and profession, can either support or hinder one's efforts to overcome social obstacles. Understanding this intricate web of identities is crucial for those seeking to drive change. Social psychology research underscores the significance of community support and collective identity in overcoming societal inertia. When individuals recognize shared experiences and goals within their communities, they can leverage collective action to break through societal boundaries more effectively.

Artificial intelligence offers fascinating potential in modeling and predicting the pathways of societal transformation, providing insights that were previously inaccessible. By analyzing extensive datasets of human behavior, AI algorithms can identify patterns and anomalies that indicate shifts in societal conventions. These models can offer predictive insights into emerging trends and potential points of resistance within societies. For example, machine learning tools can simulate scenarios where individuals successfully navigate social obstacles, providing a roadmap for fostering societal change. This advanced approach not only enhances our understanding of social dynamics but also equips individuals and organizations with the tools to anticipate and respond to shifts in community expectations.

To apply these insights practically, individuals can focus on cultivating a mindset that embraces change and uncertainty. Encouraging curiosity and experimentation in daily interactions can lead to a deeper understanding of social structures and personal roles within them. Engaging in community dialogues and fostering environments that value diverse perspectives can also enhance one's ability to challenge and redefine norms. By adopting these strategies, individuals can actively participate in the ongoing evolution of societal values, much like particles in a quantum system, constantly in motion and influencing each other in unpredictable yet transformative ways. These collective actions contribute to a more dynamic and adaptable society, capable of evolving in response to the complex challenges of the modern world.

The Influence of Entangled Identities on Social Change

Entangled identities offer a unique perspective on social change, akin to the quantum phenomenon of entanglement, where particles influence each other regardless of distance. In society, individuals often find their identities intertwined with others, creating a shared sense of purpose that can drive significant social transformations. This interconnectedness can dismantle entrenched norms by fostering a collective consciousness that transcends individual limitations. Such unity empowers groups to challenge and reshape societal paradigms, driving progress through a unified front.

Exploring the dynamics of entangled identities reveals how social movements gain momentum and achieve lasting impact. Recognizing shared experiences and aspirations, people form networks of mutual support, propelling movements forward. This shared identity becomes a powerful force that can disrupt the status quo, as seen in historical movements like civil rights and environmental advocacy. The strength of these movements often lies in their ability to unite diverse individuals under a common cause, creating a ripple effect that influences broader societal norms. Understanding entanglement in social contexts allows individuals and groups to harness this collective power to advocate for change more effectively.

Artificial Intelligence opens new avenues to analyze and predict the impact of entangled identities on social change. By leveraging sophisticated algorithms and data analysis, AI can uncover connections between individuals that might remain hidden. These insights help identify potential catalysts for social change and anticipate the trajectory of emerging movements. AI's ability to process vast amounts of information allows for a nuanced understanding of how entangled identities influence social dynamics, offering predictive models that guide strategic decision-making for activists and policymakers. As AI continues to evolve, its role in decoding social entanglement will likely become even more integral to understanding and navigating complex social landscapes.

Despite the potential of entangled identities to drive positive change, challenges and complexities must be considered. Diverse perspectives within a movement can lead to internal conflicts, hindering collective action if not managed effectively. Balancing a unified identity with respecting individual differences is crucial in sustaining momentum and achieving meaningful outcomes. Examining real-world scenarios where entangled identities have succeeded or struggled in effecting change can provide valuable lessons. By studying these case studies, one can glean insights into strategies that foster successful collaboration and pitfalls to avoid.

In contemplating the influence of entangled identities, readers are encouraged to reflect on their roles within social networks and movements. Consider the power of shared identity in your community and how it can challenge norms and foster progress. Engaging with others who share common values can amplify individual efforts and contribute to broader societal shifts. Reflect on how to identify and connect with like-minded allies and maintain cohesion within diverse groups. By exploring these questions, readers can better understand and navigate the complexities of social entanglement, ultimately contributing to a more equitable and dynamic society.

AI-Driven Models of Social Tunneling and Their Implications

AI-powered models that explore social breakthroughs offer a fascinating perspective on how people navigate and surpass cultural standards. These models utilize machine learning to uncover patterns in human behavior, resembling the concept of quantum leaps in physics, where individuals break through established societal boundaries. By sifting through extensive data from social media, forums, and digital interactions, AI can detect subtle shifts in public sentiment and individual actions that might herald a transformation in social dynamics. These insights create a predictive framework for anticipating societal evolution, providing glimpses into possible futures where unconventional behaviors become the norm.

A compelling illustration of this is the application of AI in understanding changing attitudes toward previously stigmatized issues, like mental health. By analyzing patterns in discourse and sentiment, AI can pinpoint influential figures and pivotal moments that drive shifts in perception. This analysis underscores the delicate interplay between personal initiative and collective acceptance, revealing how social breakthroughs often start with a few courageous voices challenging the status quo. These trailblazers, akin to particles in a quantum leap, find pathways through formidable barriers, gradually widening these routes for others to follow.

Beyond theoretical musings, predictive models of social breakthroughs have practical uses in marketing, public policy, and organizational change. Businesses can use these insights to predict shifts in consumer behavior, enabling more flexible and adaptive strategies. Policymakers can leverage an understanding of social breakthroughs to design more inclusive and responsive policies that address emerging societal needs. Organizations can adopt these models to create environments that encourage innovation and challenge existing norms, fostering a culture where new ideas can flourish without traditional constraints.

Despite the promise of these AI-driven models, it's essential to acknowledge the challenges and ethical considerations they pose. The reliance on data that may carry existing biases requires careful calibration and transparency in model development. Moreover, the risk of AI inadvertently reinforcing stereotypes or promoting conformity highlights the importance of diverse data sources and inclusive algorithmic design. Addressing these challenges demands a nuanced understanding of both the potential and pitfalls of AI in social analysis, ensuring its application in social breakthroughs is both ethical and effective.

To apply these insights practically, readers might consider how they can become catalysts for social breakthroughs in their communities. By embracing diverse perspectives and questioning entrenched norms, individuals can contribute to a culture of change. Engaging in conversations that challenge the status quo and supporting initiatives that promote inclusivity are actionable steps toward creating environments where social breakthroughs can thrive. Encouraging others to join this process creates a ripple effect, amplifying the impact of individual actions and facilitating broader societal transformation. Through this understanding, readers are empowered not only to grasp the mechanics of social change but to actively participate in its evolution.

The Role of AI in Predicting Social Tunneling and Behavioral Shifts

Picture a realm where the enigmatic patterns of human behavior are deciphered with the precision akin to the principles of quantum physics. At the core of this exploration is the captivating notion of social navigation—where individuals transcend the gravitational pull of societal conventions, breaking through unseen boundaries much like particles making unexpected leaps in quantum transitions. This phenomenon, though abstract, is far from arbitrary. With the rise of artificial intelligence, we stand on the verge of a revolution in understanding these behavioral transformations. AI, with its exceptional capability to analyze immense volumes of data, promises to unravel the intricate signals that precede these social breakthroughs. Just as quantum physics uncovers the hidden movements of particles, AI reveals patterns within the intricate fabric of human interaction, offering insights into the forces that propel social evolution.

Leveraging AI for this purpose revolutionizes the field of social analysis, shedding light on the pathways of social navigation with unmatched clarity. Machine learning algorithms, acting like keen observers, learn to identify the markers that signal these shifts, while neural networks predict the transitions with a foresight previously thought unattainable. Inspired by quantum mechanics, quantum-inspired algorithms further refine these predictive models, enhancing their precision and breadth. Incorporating these insights into real-time social dynamics forecasting promises not just to forecast, but to comprehend the underlying causes of these shifts. This chapter embarks on a journey through these advancements, illustrating how AI not only anticipates but also deepens our understanding of the social world, unlocking new possibilities for both individual and collective progress.

Utilizing Machine Learning to Identify Patterns in Social Tunneling

Using machine learning, researchers and social scientists explore the complexities of social evolution, much like quantum particles overcoming barriers through tunneling. By examining large datasets from social media, surveys, and real-world interactions, machine learning algorithms detect the subtle signals that foreshadow changes in social norms and behaviors. These tools excel at uncovering intricate patterns and correlations that often go unnoticed by humans, providing a deeper understanding of how societal barriers are transcended. For example, by analyzing historical data on social movements and individual actions, machine learning can identify pivotal moments when collective attitudes began to shift, offering insights into the conditions that might lead to similar transformations in the future.

One promising application of machine learning is its ability to identify emerging patterns using unsupervised learning techniques. Unlike traditional methods that depend on predefined categories, unsupervised learning uncovers novel patterns and groupings within data, illuminating the spontaneous and unpredictable nature of social evolution. By clustering individuals based on shared characteristics or behaviors, these techniques reveal the underlying structures in social networks that either facilitate or impede the breaking of norms. This approach not only deepens our understanding of social dynamics but also equips policymakers and leaders with the knowledge to create environments that foster positive social change by pinpointing leverage points within these networks.

Additionally, the integration of reinforcement learning offers a dynamic method for modeling social evolution. This technique simulates decision-making processes within a social context, allowing for the exploration of various scenarios and their potential outcomes. By iteratively learning from interactions and feedback, reinforcement learning models can predict how individuals might navigate and overcome societal barriers over time. These simulations inform strategies to encourage innovation, acceptance of diversity, and the dismantling of outdated social constructs, turning theoretical insights into practical tools for understanding and influencing social progress.

Recent advancements in explainable AI are crucial in clarifying the machine learning models used to study social evolution. These developments allow researchers to interpret the often opaque decision-making processes of AI, providing transparency and accountability in the analysis of social data. Understanding the reasoning behind AI-driven predictions gives stakeholders confidence in the insights provided, enabling informed decisions based on these findings. This transparency is essential, as it permits critical examination of the assumptions and biases that may be embedded within the data and models, ensuring that conclusions are robust and ethically sound.

The application of machine learning in identifying patterns of social evolution not only enhances our comprehension of human behavior but also offers actionable insights for promoting societal progress. By recognizing early indicators of social change, individuals and communities can better navigate and influence these shifts, fostering a more inclusive and dynamic social environment. As machine learning continues to advance, its role in the social sciences holds the potential to revolutionize our understanding and engagement with the world, encouraging a future where societal barriers are not just crossed but redefined.

Leveraging Neural Networks to Anticipate Behavioral Transitions

Neural networks have become indispensable tools for decoding the complexities of human behavior, offering deep insights into predicting changes in societal norms and personal actions. These advanced systems, modeled after the human brain's structure, excel at uncovering intricate patterns within extensive datasets. By analyzing historical social data, neural networks can pinpoint subtle signs of upcoming behavioral shifts, similar to how they are utilized in finance to forecast market trends. This ability enables a refined understanding of when individuals or groups might diverge from established conventions, offering a glimpse into the mechanics of societal breakthroughs, where people transcend cultural boundaries.

Recent progress in neural network architecture, especially the advent of deep learning models, has markedly improved their predictive capabilities. These models are adept at processing layers of information, capturing intricate dependencies and relationships that simpler models might miss. For example, researchers have used these networks to study social media interactions, revealing hidden societal trends before they surface in offline behaviors. This application highlights the potential of neural networks to foresee shifts in public sentiment or collective actions, such as the rapid rise of social movements or changes in consumer preferences, by detecting patterns that are not immediately noticeable to the human eye.

Incorporating neural networks into the study of social behaviors also challenges traditional assumptions about the predictability of human actions. Whereas classical models often depend on static assumptions, neural networks embrace the dynamic and fluid nature of human societies, adapting their predictions in real time as new data emerges. This adaptability mirrors the principles of quantum mechanics, where uncertainty and probability are essential elements of understanding phenomena. By adopting a quantum-inspired approach, neural networks offer a more flexible framework for predicting societal breakthroughs, accommodating the inherent unpredictability of human interactions.

To fully leverage the potential of neural networks in predicting behavioral transitions, interdisciplinary collaboration is crucial. Bringing together expertise from computer science, sociology, and cognitive psychology can enrich the development of algorithms that are not only technically proficient but also contextually relevant. This cross-disciplinary exchange fosters innovative problem-solving approaches, such as incorporating ethical considerations into predictive models to ensure AI-driven insights promote positive social outcomes. Engaging diverse perspectives also helps mitigate biases that can arise from homogeneous data sources, enhancing the fairness and accuracy of predictions.

Neural networks' capacity to forecast social transformations invites reflection on the broader implications of AI in society. As these technologies become more integrated into decision-making processes, questions arise about their role in influencing human behavior. For instance, how should insights from neural networks be applied in policymaking, urban planning, or conflict resolution? By encouraging readers to ponder these questions, this exploration of neural networks not only highlights their technical capabilities but also stimulates critical thinking about their ethical and societal dimensions. The fusion of AI and social sciences holds significant potential for understanding and shaping the future, urging readers to consider how best to navigate this evolving landscape.

Applying Quantum-Inspired Algorithms for Enhanced Social Predictive Models

Quantum-inspired algorithms are opening new horizons in social predictive modeling by offering innovative perspectives on human behavior's complexity. Inspired by quantum mechanics principles like superposition and entanglement, these algorithms handle data in ways traditional methods cannot. They simulate multiple scenarios simultaneously, providing a comprehensive understanding of social behaviors akin to observing all potential states at once. This is particularly useful in identifying social breakthroughs where individuals or groups challenge norms to drive change. As the intersection of quantum computing and AI continues to be explored, the ability to predict behavioral shifts becomes more achievable, presenting a cutting-edge toolkit for sociologists and data scientists.

In predictive analytics, the flexibility of quantum-inspired algorithms is noteworthy. While traditional models often falter with the uncertainty and variability inherent in human behavior, quantum-inspired approaches embrace these complexities. By incorporating elements of uncertainty and probability, they reflect the fluid nature of social dynamics, enabling more nuanced predictions. This adaptability is critical where behavior stems from complex influences rather than linear trends. For example, the sudden emergence of grassroots movements can defy established patterns; quantum-inspired modeling can offer insights into their societal impact, identifying potential catalysts and forecasting their paths.

A fascinating application of these algorithms is their ability to weave together diverse data sources, creating a coherent narrative from the chaotic landscape of human interaction. By integrating information from social media, economic indicators, and cultural trends, these models construct a multi-dimensional view of societal shifts. This holistic approach is like assembling a jigsaw puzzle, with each piece representing a different aspect of human behavior. As patterns become clear, these models can pinpoint where entrenched norms might begin to dissolve, paving the way for significant change. This capability not only aids in understanding current dynamics but also in predicting future societal landscapes with greater accuracy.

Exploring quantum-inspired algorithms in social contexts also prompts a reconsideration of ethical issues and AI's role in shaping human interactions. As these models grow more adept at predicting and potentially influencing social behavior, questions about autonomy and free will emerge. How do we balance the benefits of precise predictions with the risk of unintended social steering? Addressing these concerns requires a multidisciplinary approach, drawing from philosophy, ethics, and technology. By fostering dialogue across these fields, we can better navigate the complexities of using quantum-inspired insights responsibly, ensuring they enhance rather than diminish human agency.

For professionals in the field, the practical implications of quantum-inspired social models are significant. Researchers and practitioners can leverage these algorithms to not only forecast societal trends but also design interventions that promote positive change. By identifying potential points of social breakthroughs, policymakers and community leaders can proactively address emerging issues, fostering environments that support innovation and adaptability. As AI and quantum computing continue to evolve, staying informed about these advancements is crucial for those aiming to use technology in meaningful, socially responsible ways. This journey into the quantum-inspired realm promises to reshape our understanding of society, offering tools that are as transformative as they are insightful.

Integrating AI Insights into Real-Time Social Dynamics Forecasting

Artificial intelligence has transformed our capacity to predict and understand social dynamics in real-time by leveraging extensive datasets and discerning intricate patterns. Machine learning algorithms, especially those drawing inspiration from quantum principles, provide advanced means for analyzing social behaviors instantly. These algorithms can swiftly and accurately process complex data, detecting subtle changes that may indicate shifts in societal norms or the emergence of new trends. By integrating variables from diverse sources like social media platforms, economic metrics, and cultural happenings, AI can deliver predictions that are both timely and contextually nuanced, offering a more dynamic grasp of societal transformations and behavioral shifts.

Embedding AI insights into real-time social forecasting entails more than mere data aggregation; it demands an in-depth comprehension of human psychology and societal frameworks. Advanced neural networks, for instance, can replicate human decision-making processes, yielding insights into how groups might respond to specific stimuli or environmental changes. These models can anticipate how societal pressures could affect individual actions, providing a glimpse into potential paths of social evolution. Through scenario simulations, AI can propose proactive strategies for overcoming social obstacles, enabling individuals and organizations to foresee and adapt to change with agility.

The fusion of AI insights into social dynamics forecasting also creates opportunities for personal and organizational development. For individuals, understanding the potential for societal transformations can lead to more informed decisions regarding career choices, social interactions, and personal relationships. Organizations can use these forecasts to refine their strategies, ensuring they stay responsive to evolving social landscapes. By anticipating behavioral shifts, businesses can innovate more effectively, aligning their products and services with the emerging needs and values of their consumers. This capability not only enhances competitiveness but also fosters a more resilient societal framework that can accommodate change.

Furthermore, the use of AI in predicting social dynamics extends beyond macro-level analysis. On a micro scale, it offers significant implications for personal growth and interpersonal relationships. By examining patterns in individual behavior and social interactions, AI can provide personalized insights that promote self-awareness and growth. These insights can assist individuals in identifying their own social hurdles and developing strategies to overcome them, fostering a sense of empowerment and agency. For communities, AI-driven insights can facilitate more inclusive and harmonious interactions, guiding efforts to bridge cultural divides and promote social cohesion.

As AI continues to advance, its role in real-time social dynamics forecasting will undoubtedly broaden, presenting both challenges and opportunities. The ethical implications of using AI to predict and influence human behavior must be carefully considered, ensuring that such technologies are deployed responsibly and with respect for individual autonomy. Thought-provoking questions arise: How can we balance the benefits of predictive analytics with the need for privacy and consent? What frameworks are necessary to ensure that AI-driven insights are used ethically and equitably? By addressing these questions, society can harness the transformative potential of AI to foster a more adaptive and interconnected world, where transcending social obstacles becomes a collective journey of discovery and growth.

Reflecting on the essence of this chapter, the intriguing connections between quantum mechanics and societal dynamics offer fresh perspectives on how people navigate and reshape cultural boundaries. By using quantum transitions as a metaphor for overcoming hurdles, readers can gain a deeper understanding of the innovative strategies individuals employ to defy norms and spark transformation. The potential of AI to anticipate these shifts in behavior highlights the groundbreaking role of technology in deciphering intricate social patterns. This exploration encourages a reimagining of human agency and resilience within society. As the narrative progresses, consider the influence of these quantum ideas not just on personal actions but also on the collective development of communities. This journey illustrates the delicate dance between physics and human conduct, prompting a thoughtful reflection on the unseen forces that mold our social environments. As we move forward to the next chapter, we should ponder what additional insights from quantum mechanics might further illuminate the nuances of human connections and societal evolution.

Chapter Six

Group Dynamics And Quantum Coherence

Picture yourself in the heart of a bustling train station during rush hour. The throng of commuters moves with a rhythm that appears chaotic at first glance, yet reveals a hidden order upon closer inspection. Like an intricate dance, individuals navigate through the crowd with an innate awareness of their surroundings, bound by invisible connections. This familiar yet captivating scene mirrors the complexity of human interactions. While it might seem random, a deeper look unveils a harmony akin to the surprising coherence found in the world of particles. In this chapter, we embark on a journey to explore the parallels between these realms, shedding light on how principles of quantum harmony can illuminate the intricacies of human collective behavior.

As we delve deeper, the concept of harmony will transcend its scientific roots, becoming a lens through which we view social unity. Just as quantum harmony preserves the delicate balance of particles, social unity holds groups together, nurturing a shared sense of purpose and direction. Imagine a choir where each voice contributes to a perfect blend, or a sports team that moves with seamless coordination. These instances of collective alignment resonate with quantum principles, offering insights into the unseen forces that keep communities in sync, even amidst the chaos of daily life.

Our exploration will venture into the interplay between harmony and disruption, examining how collectives either maintain their unity or descend into disorder. Through the perspective of physics, we gain a fresh understanding of collective action, uncovering the intricate balance that sustains societal structures. This chapter aims to illuminate the invisible threads woven into human interactions, drawing parallels between the subatomic and the social. By grasping these connections, we unlock the potential to perceive our social world anew, appreciating the hidden symmetries that influence our lives.

Quantum Coherence: A Model for Understanding Unified Groups

Imagine a world where the harmony within a group mirrors the elegant choreography of particles in a state of subatomic synchronization. This intriguing comparison invites us to explore how individual actions unite to form a cohesive whole, much like particles acting in unison despite their distinct natures. In human interactions, this unity is evident in successful teams, communities, and organizations, where interconnected individuals create a shared purpose and identity, fostering innovation and collective achievement. The concept of particle coherence offers a fascinating framework for understanding how diverse components can blend seamlessly, erasing the boundaries between individual and collective identities and unveiling the forces that tie them together.

Embarking on this exploration reveals the foundational role of superposition in shaping group identities. Just as particles simultaneously exist in various states, individuals contribute to a collective identity that embraces a spectrum of perspectives and possibilities. This fluidity promotes adaptability and resilience—vital traits for thriving in ever-changing social environments. Entanglement enriches this narrative, showing how one person's decisions can resonate through a community, profoundly influencing collective choices. The observer effect, often seen as a curious aspect of physics, underscores the power of perception in fostering unity or discord. Ultimately, the resonance from these interactions can be harnessed to amplify team synergy, transforming a collection of individuals into a dynamic force for collective action. This particle-inspired perspective offers a transformative view of group dynamics, encouraging us to appreciate the intricate dance of harmony defining human connections.

The Role of Quantum Superposition in Group Identity Formation

Subatomic particles, with their enigmatic and ever-changing nature, offer a fascinating analogy for understanding how collective identities form and transform. The principle of quantum superposition, where particles exist in multiple states until observed, reflects the fluid nature of forming group identities. As individuals join a collective, they bring a variety of perspectives, skills, and experiences. These elements coexist, much like states in superposition, until the group collectively shapes its identity. This process remains dynamic, open to change and adaptation through the contributions and interactions of its members. Recent studies in social dynamics reveal that group identity is not a singular, fixed entity but a spectrum of possibilities that shifts with context and interactions. This concept aligns with quantum superposition, where identity is not limited to one state but is a blend of several potential states. For instance, a sports team might simultaneously embrace competitiveness, supportiveness, and community orientation, embodying these traits based on the situation. This multifaceted nature allows collectives to adapt to various challenges, fostering resilience and innovation.

Consider the workplace, where teams tackle complex projects and navigate diverse personalities. Here, each member's potential contributions exist in a state of superposition, with the group's identity solidifying only through decisions and actions. By embracing this fluidity, teams can leverage the full spectrum of their members' potential, fostering an environment rich in creativity and collaboration. Encouraging multiple viewpoints to coexist without immediate resolution can lead to more innovative solutions and a more cohesive group identity.

The application of quantum superposition in social contexts also highlights the importance of flexible leadership and decision-making. Leaders who appreciate the value of maintaining a superpositional state within their teams can better guide group dynamics, allowing for a more organic and inclusive identity development. This approach can also mitigate conflicts by viewing diverse perspectives as strengths rather than points of contention. By fostering an environment where group identity is seen as evolving, leaders can help their teams navigate complex social landscapes with greater agility and effectiveness.

To explore these ideas further, consider how teams might consciously adopt practices that encourage superposition-like states. This could include creating spaces for open dialogue, encouraging experimentation, and valuing diverse contributions without rushing to form a singular identity. Such practices not only enhance group cohesion but also prepare teams to face future challenges with adaptability and collective insight. By recognizing and embracing the parallels between quantum superposition and group identity formation, individuals and organizations can cultivate more dynamic and resilient social structures.

Entanglement and Its Influence on Collective Decision-Making

In the fascinating world of quantum physics, entanglement describes a mysterious link between particles, where the condition of one instantaneously affects another, despite the distance between them. This concept intriguingly mirrors the interconnectedness seen in group decision-making. When individuals unite, their thoughts and actions can intertwine, transcending personal intentions and forming a collective force that steers group decisions. Much like entangled particles share a state, members of a tightly-knit team often experience a synchronization of goals and motivations, leading to decisions that reflect a collective mindset rather than isolated preferences. This interconnected nature of group dynamics can be leveraged to promote collaboration and innovation, encouraging the emergence of ideas greater than the sum of their parts.

Recent findings in social psychology and neuroscience highlight the effect of entanglement in group settings. Research shows that when individuals view themselves as part of a cohesive unit, their neural patterns align, echoing the phenomenon of quantum entanglement. This alignment transcends metaphor; it influences how information is processed and decisions are reached. As team members engage in discussions, shared experiences and emotional bonds contribute to a unified decision-making process. This synergy bolsters the group's ability to address complex problems, as diverse perspectives and expertise are seamlessly integrated into a coherent strategy. These insights suggest that nurturing strong interpersonal connections within teams can replicate the advantages of quantum entanglement, leading to more effective and harmonious decision-making.

A captivating aspect of entanglement in group decision-making is the potential to achieve consensus without direct communication. In the quantum world, particles affect each other instantaneously, bypassing traditional communication channels. Similarly, in human groups, a well-established understanding and shared vision can create an implicit agreement among members, often termed "group intuition." This phenomenon shines in high-performing teams, where members anticipate each other's needs and actions, reducing the need for explicit directives. Organizations can cultivate this intuitive synergy by fostering a culture of trust and openness, enabling members to internalize and align with the team's objectives, much like entangled particles naturally align their states.

Exploring the quantum principle of entanglement in the context of collective decision-making offers valuable insights into enhancing organizational effectiveness. One actionable insight is the significance of fostering a shared identity and purpose within groups. This can be achieved through team-building activities, clear communication of the group's mission, and recognizing individual contributions while emphasizing collective achievements. Leaders play a crucial role in creating environments that encourage open dialogue and mutual respect, paving the way for entangled connections to thrive. By doing so, they enable teams to operate with a level of harmony and efficiency akin to the quantum realm.

Imagining the future of group decision-making through the lens of quantum entanglement invites us to consider its potential to revolutionize organizational dynamics. Could principles of quantum physics inspire new technologies or methods that enhance collaborative efforts? As artificial intelligence evolves, exploring its role in facilitating and analyzing these interconnected interactions could unlock new dimensions of group performance. Embracing the interconnectedness of individuals within a team may not only elevate decision-making processes but also inspire a more holistic and integrated approach to problem-solving, ultimately transforming how we collaborate in a rapidly changing world.

The Observer Effect's Impact on Group Cohesion and Performance

The intricate interplay of group dynamics brings forth the observer effect, a crucial concept illustrating how external observation can alter a collective's harmony and performance. Originating from quantum physics, this idea posits that observation can influence a system's state, paralleling how groups react when they are watched. Awareness of evaluation often prompts team members to modify their behavior, sometimes enhancing unity or, conversely, inciting discord. The expectation of judgment can drive individuals to adhere more closely to group norms or exacerbate tensions as they negotiate personal and shared identities.

In high-performing teams, awareness of the observer effect can be strategically advantageous. Imagine a project team aware of upper management's scrutiny, consciously using this awareness to increase productivity and unity. By cultivating transparency and open communication, the team can manage stress linked to observation, channeling pressure into a cohesive force that boosts performance. This deliberate use of external observation transforms a potential disruption into a catalyst for synchronization and synergy.

Recent organizational behavior studies reveal that the observer effect acts both as a mirror and a mold, reflecting and shaping group dynamics. Research from leading business schools shows that anticipated evaluation often leads to increased conformity and a stronger shared purpose. While beneficial in scenarios requiring collective action, this phenomenon can also hinder creativity and innovation if not carefully managed. By understanding these subtleties, leaders can create environments that leverage the observer effect to foster solidarity while preserving individuality.

To navigate the complexities introduced by the observer effect, groups can adopt strategies that enhance resilience and adaptability. Promoting reflection and flexibility helps members remain responsive to external pressures. Implementing regular feedback loops focused on growth rather than judgment can alleviate anxiety and reinforce a culture of continuous improvement. By reframing observation as an opportunity instead of a threat, groups can cultivate resilient unity that thrives under scrutiny.

At the intersection of physics and social psychology, the observer effect prompts contemplation on how awareness of being observed influences interactions. Insights from both fields offer opportunities to rethink how groups can function at their best. Encouraging teams to view observation as a dynamic aspect of their existence opens avenues for growth and creativity. Thought-provoking questions, such as how external perceptions can be harnessed to foster innovation or maintain group integrity, invite further exploration and application of these principles in diverse settings.

Harnessing Quantum Resonance for Enhanced Team Synergy

Quantum resonance serves as an intriguing metaphor for team dynamics, illustrating how individuals can synchronize to form a powerful collective force. Imagine a team operating like particles in resonance, each member attuned to the others, creating synergy that elevates their collective output beyond individual contributions. This alignment allows teams to achieve remarkable results by complementing each member's strengths and weaknesses, much like particles enhancing a wave through resonance. This metaphor underscores the significance of collaboration and coordination, where the team becomes more impactful than the sum of its parts. Such harmony enables teams to tackle complex challenges deftly, make swift decisions, and maintain a unified vision.

Recent studies on quantum-inspired organizational behavior models reveal the transformative power of resonance in boosting team dynamics. Applying quantum coherence principles, organizations can cultivate environments that nurture creativity and innovation. For example, companies might use AI-driven tools to analyze communication patterns, identifying areas of resonance within teams. These insights can guide interventions that enhance alignment and cooperation, ensuring everyone is in sync with the team's rhythm. Such strategies not only increase productivity but also boost job satisfaction, as individuals feel more connected and engaged with their colleagues.

Harnessing quantum resonance in practice requires more than aligning tasks and goals; it demands fostering a culture that prioritizes emotional intelligence and empathy. Encouraging teams to explore diverse viewpoints and engage in open communication leads to a state of coherence where ideas flow effortlessly, and innovative solutions arise naturally. This environment, akin to a fluid quantum state, supports adaptability and resilience—key traits in today's rapidly evolving world. Leaders play a vital role in nurturing this atmosphere, guiding teams toward resonance without enforcing rigid structures.

The concept of quantum resonance also challenges traditional hierarchical teamwork models. Instead of viewing leadership as a top-down directive, it suggests a more distributed approach with fluid leadership roles that adapt to context. This flexibility mirrors the probabilistic nature of quantum systems, where outcomes emerge from interactions and relationships rather than being predetermined. By embracing this mindset, organizations can create a more inclusive and empowering environment, allowing every team member to lead and contribute meaningfully, thus enhancing the group's overall unity and synergy.

Envisioning teams as systems of quantum resonance prompts critical questions about collaboration and the factors that drive successful teamwork. How can organizations create conditions for spontaneous coherence among diverse individuals? What role does technology play in facilitating or obstructing this process? By exploring these questions, we can develop actionable strategies to cultivate resonance in teams, leading to enhanced synergy and improved organizational outcomes. This exploration deepens our understanding of human dynamics and provides practical insights for building more harmonious and effective teams in any setting.

Social Coherence: What Keeps Groups in Sync

Examining human dynamics reveals that, much like the delicate interplay of particles in a subatomic system, human collectives display their own form of harmony—a subtle yet powerful force that keeps them in sync. Imagine a school of fish gliding effortlessly through water, each individual moving as part of a larger, unified entity. This synchronized behavior emerges not from a central figure but from a shared understanding and connection among the fish. In human contexts, social harmony operates similarly, binding individuals into cohesive communities. This enigmatic force, reminiscent of subatomic unity, allows groups to function as cohesive units, adeptly navigating complex social landscapes. At its essence, social harmony transforms a mere gathering of individuals into a potent collective identity, capable of influencing the surrounding world.

To grasp the mechanics of this phenomenon, we can draw fascinating parallels with the realm of particle physics. Quantum entanglement offers insights into how collective identities form, linking individuals through invisible strands of shared values and aspirations. Just as particles can exist in multiple states simultaneously, communities often navigate a state of flux during decision-making. The observer effect also plays a role, subtly shifting group behaviors simply through the awareness of being observed. Furthermore, much like how quantum tunneling allows particles to bypass barriers, communities frequently find ways to overcome social obstacles, penetrating what seems impenetrable. By exploring these connections, we unravel a rich tapestry of understanding about what truly keeps communities in sync, offering a deeper appreciation for the unseen forces at play in our social interactions.

Quantum Entanglement and Collective Identity Formation

Quantum entanglement provides an intriguing perspective for examining how collective identity emerges within groups. In physics, entangled particles maintain a connection that allows the state of one to instantly affect the other, regardless of distance. This phenomenon mirrors social dynamics, where individuals within a unified community often share an identity that transcends personal differences. Such a collective identity functions like entanglement, enabling a group to act with a common purpose despite the varied views of its members. Recent research in social psychology indicates that when individuals see themselves as interconnected, they are more inclined to uphold the group's values and goals, illustrating a kind of social harmony.

The application of quantum entanglement to the formation of group identity challenges conventional views of individuality versus collectivism. Rather than viewing these as conflicting forces, entanglement suggests a dynamic interaction where personal identities can coexist with and even strengthen group unity. This model is evident in innovative organizations that thrive on collaboration, where employees are encouraged to maintain their unique identities while contributing to a shared mission. Such environments nurture creativity and innovation, as the interconnected nature of their social structure facilitates seamless cooperation and swift idea exchange. This synergy underscores the potential for quantum-inspired frameworks to influence organizational growth and team dynamics.

Considering collective identity as a form of social entanglement offers a fresh perspective on group resilience. Just as entangled particles show resilience against external disruptions, groups with a strong collective identity can endure challenges and adapt to changing conditions. For example, communities facing adversity often rely on shared values and a sense of belonging to navigate crises, emerging more robust and cohesive. This resilience is not merely the result of shared experiences but a testament to the entangled bonds that connect group members, enabling them to support one another and maintain social unity.

In exploring the implications of social entanglement, it's crucial to consider how technology, particularly artificial intelligence, can enhance or disrupt these connections. AI tools capable of analyzing complex social networks can pinpoint patterns of entanglement and propose strategies for boosting group unity. However, technology also risks undermining these bonds by emphasizing efficiency over meaningful interaction. Balancing technological advancement with the preservation of human connection is vital for fostering social structures that remain adaptable and resilient. Engaging with these considerations invites a critical examination of the role technology plays in shaping collective identities.

To apply these insights practically, consider the role of leadership in fostering social entanglement. Leaders can nurture a sense of collective identity by promoting inclusive environments where diverse perspectives are valued and integrated into the group's mission. Encouraging open communication, acknowledging individual contributions, and aligning group objectives with members' values can enhance social unity and ensure that interconnected bonds remain strong. By embracing the principles of quantum entanglement, leaders can guide their teams toward a harmonious balance between individuality and unity, paving the way for more innovative and resilient communities.

Superposition States in Group Decision-Making

In the complex realm of collective decision-making, the principle of superposition presents a fascinating model for how groups can navigate uncertainty and intricate challenges. Originating from quantum physics, superposition allows particles to exist in multiple states at once. When applied to social settings, this concept helps explain how teams can handle decisions involving various possibilities and potential outcomes. Much like particles that remain in several states until measured, teams can simultaneously consider diverse viewpoints and options, which fosters flexibility and adaptability. This 'decision superposition' approach enables groups to discover innovative solutions and explore paths that might be missed if decisions were made in a straightforward manner.

Recent studies in cognitive science and organizational behavior highlight the benefits of adopting a superpositional mindset. Research shows that groups with high cognitive diversity—where members bring different backgrounds, expertise, and perspectives—are better equipped to navigate these fluid decision-making states. Such diversity creates a breeding ground for new ideas, enhancing the group's ability to integrate information from various sources and develop unique solutions. Practically, cultivating an environment that not only welcomes but actively encourages diverse opinions can lead to more robust decision-making processes that thrive amid complexity.

Emerging technologies, especially AI and machine learning, provide powerful tools to aid groups in maintaining superpositional states. By analyzing extensive datasets and simulating outcomes, AI can offer insights that are hard to achieve using traditional methods. This technology enables groups to visualize the intricate nature of their decisions, exploring different scenarios without quickly settling on a single choice. AI-driven decision-support systems thus act as catalysts for maintaining superposition, guiding teams through a maze of options with increased clarity and precision.

To fully leverage superposition in decision-making, groups need to foster a culture of openness and experimentation. Encouraging members to embrace uncertainty and resist hasty conclusions can transform decision-making from a rigid process into a dynamic exploration. Activities like envisioning multiple future scenarios or role-playing different perspectives can serve as effective tools to prolong this exploratory phase. By doing so, groups can unlock a range of opportunities beyond the immediate horizon.

Considering the broader implications of superposition in group dynamics, one might wonder how these principles could be applied to societal challenges. Can communities adopt a superpositional approach to address complex issues like climate change or social inequality, where multiple solutions must coexist until the best path emerges? These questions prompt a rethinking of collective decision-making, encouraging a shift from binary thinking to a more nuanced appreciation of possibilities. This paradigm enriches the decision-making process and nurtures a culture of innovation and adaptability, essential for navigating the ever-changing landscape of human interaction.

The Observer Effect and Its Impact on Group Behaviors

The observer effect, a fundamental principle of quantum physics, illustrates how observation can significantly alter the behavior of particles. This intriguing concept finds a parallel in social dynamics, where merely being observed can influence how individuals act within groups. In various social settings, people often adjust their behavior when they know they are being watched by peers, authorities, or even through self-monitoring. Much like the quantum observer effect, this phenomenon suggests that observation itself can either unify or disrupt group synergy, impacting decision-making and collective actions. Recent social psychology research shows that individuals tend to conform to perceived standards when they are observed, underscoring the influence of social contexts on behavior.

In organizational environments, the observer effect is evident. Picture a workplace where employee performance shifts when a supervisor is nearby. Awareness of being watched can boost productivity, encourage adherence to norms, or suppress creativity due to fear of judgment. This dynamic highlights the observer's impact on group behavior and offers insights into effective team management. Leaders can use this understanding to cultivate environments that promote positive behaviors, fostering a culture of openness and innovation rather than compliance or fear.

The observer effect extends into the digital world, where online interactions and social media offer new observation forms. Users often tailor their online personas, knowing they are being scrutinized by a wide audience, including friends and potential employers. This digital observer effect influences how people present themselves and interact, affecting group unity in virtual communities. Understanding these dynamics can help in designing digital platforms that encourage genuine interactions and lessen the constant observation pressure.

Consider a thought experiment involving group decision-making, contrasting scenarios where participants remain anonymous with those where they are identifiable. The outcomes often vary, with anonymity promoting honesty and risk-taking, while identifiability can lead to conformity and risk aversion. This mirrors the quantum observer effect, where observation changes the system's state. Creating environments that balance transparency with anonymity can result in more authentic and effective group interactions, emphasizing the nuanced role of observation in social synchronization.

The observer effect in group behaviors compels us to rethink how observation shapes social interactions. By acknowledging the intricate interplay between observation and action, individuals and organizations can intentionally foster environments that support authentic interactions and collective well-being. This understanding equips leaders with tools to navigate complex social landscapes, ensuring that the observer's role enhances rather than hinders group unity and collective advancement.

Quantum Tunneling and Breaking Through Social Barriers

Quantum tunneling, where particles traverse energy barriers they seemingly can't, serves as an apt metaphor for groups overcoming societal obstacles. In human interactions, these barriers manifest as cultural biases, hierarchical systems, or rigid norms. Much like particles using quantum tunneling to navigate energy fields, groups can leverage collective determination and innovative strategies to overcome seemingly insurmountable challenges. This process highlights the significance of a shared mission and unified identity, urging individuals to pool their efforts and resources toward a common goal.

A vivid example of this concept is seen in global civil rights movements, where marginalized communities have historically confronted formidable societal barriers. By uniting around a cohesive vision and taking collective action, these groups have successfully challenged and reformed deeply rooted social norms. Here, quantum tunneling symbolizes the seemingly impossible yet achievable transformations inspired by such movements. The strength of collective action lies in its ability to forge new pathways that were previously unseen or thought unattainable, akin to particles tunneling through barriers.

Recent research in social dynamics underscores technology and communication's role in facilitating this social tunneling. Digital platforms enable individuals from diverse backgrounds to connect, share ideas, and organize, effectively lowering the activation energy required for collective action. This digital interconnectedness mirrors the non-locality of quantum particles, allowing groups to coordinate across vast distances. The rise of social media campaigns and digital activism illustrates how modern technology can drive societal change, breaching barriers that once seemed impenetrable.

Creating environments where such breakthroughs occur requires leaders and change agents to foster trust and innovation. They must encourage members to think beyond conventional limits by promoting a culture that values diverse perspectives and creative problem-solving. This approach can stimulate significant social change, facilitating quantum leaps that transcend existing barriers and envision new frameworks for collaboration serving collective well-being.

To apply these insights practically, individuals and groups should identify the social barriers they face and strategically leverage their unique strengths and resources to address them. Encouraging open dialogue, fostering inclusivity, and embracing technological tools are actionable steps in this process. By adopting a mindset that views barriers as opportunities for innovation and growth, groups can achieve remarkable progress, much like particles finding new paths through the quantum landscape. As readers reflect on these ideas, they might consider which barriers in their lives or communities could be reimagined as gateways to transformative change.

How Collective Action Mirrors Quantum Coherence and Decoherence

Invisible forces weave through both the subatomic realm and human society, revealing striking parallels in how they orchestrate collective actions. Picture a flock of birds in perfect sync, each one subtly adjusting to the others—a scene reminiscent of particle coherence in physics. In this phenomenon, particles operate in unison, maintaining a shared state that mirrors a well-coordinated community. This synchronization in human behavior, akin to the dance of particles, prompts intriguing questions about the forces that align our actions and choices. Exploring this connection uncovers how social structures achieve unity and how groups, despite their complexity, often reach a harmonious state.

Yet, much like the temporary nature of particle harmony, unity within human collectives can dissipate, leading to fragmentation and discord. This shift is akin to decoherence, where particles lose alignment, resulting in apparent chaos. In the social sphere, such transitions often signal pivotal changes, whether in small communities or large movements. Delving deeper into these parallels reveals that collective decision-making processes often embody harmony, though the delicate balance can shift into a state of disarray where unified action dissolves into individual pursuits. Understanding these dynamics enriches our grasp of social phenomena and highlights the complex interaction between unity and fragmentation in shaping human interactions.

The Role of Quantum Coherence in Synchronizing Group Behaviors

Particles entering a state of unified harmony, known as quantum coherence, offer a captivating perspective on the synchronization of collective behaviors. In social settings, this phenomenon manifests when individuals align their actions, thoughts, and intentions, forming a cohesive unit that operates with impressive fluidity. Just as coherent particles exhibit a collective identity, human groups often achieve synergy, where individual efforts seamlessly blend toward a common purpose. This synchronization emerges as more than the sum of its parts, transforming the group into an effective and resilient entity.

Recent progress in social neuroscience has begun to uncover the neural basis of this collective synchrony, highlighting how brain waves can synchronize during group activities. This alignment fosters a shared mental landscape, enhancing empathy and understanding among participants. Studies using EEG technology have shown that synchronized brainwave patterns boost cooperation, mirroring how coherence in physics leads particles to act in unison. These discoveries underscore the importance of shared attention and intention as the foundation for cohesive interactions and decision-making.

Examining group coherence also brings to mind "swarm intelligence" observed in nature. Flocks of birds, schools of fish, and ant colonies display coordinated movements that seem almost choreographed. These natural events resemble quantum coherence, as each creature adjusts its behavior based on neighbors, creating complex, synchronized patterns without central direction. Human groups can similarly tap into this collective intelligence, adapting to changing environments by maintaining a unified structure—a crucial skill in dynamic settings where flexibility and quick response are key.

The idea of coherence in group behaviors also prompts reflection on how social networks enable synchronization. In today's hyper-connected world, digital platforms facilitate instant communication and collaboration, often leading to the rapid convergence of ideas and movements. Online communities can reach a level of unity similar to a quantum state, where collective influence far exceeds individual contributions. Understanding the parallels between coherence and digital dynamics offers valuable insights into nurturing stronger, more resilient communities capable of meaningful change.

To apply these concepts practically, it's vital to foster environments that promote harmony. Leaders can cultivate this by encouraging open communication, mutual respect, and a shared vision, allowing groups to achieve synergy. Embracing diverse perspectives enriches the collective knowledge base, boosting the group's ability to adapt and innovate. By recognizing and leveraging the parallels between coherence and group dynamics, individuals and organizations can unlock new collaboration potential and achievement, turning abstract ideas into tangible societal benefits.

Understanding Social Fragmentation Through Quantum Decoherence

The complex interplay of human collectives can be compared to the phenomenon known as quantum decoherence, where initially unified systems begin to fragment. Similar to how quantum entities lose their cohesion when interacting with external environments, social collectives can disintegrate under external pressures and internal discord. Emerging research in social neuroscience indicates that group harmony can be disrupted as individuals pursue personal interests over shared objectives. This shift can be viewed as a form of decoherence, where a once-unified group starts to break apart. By examining these phenomena through the lens of quantum physics, we gain a deeper understanding of the delicate balance necessary to maintain social unity.

Recent studies highlight the significance of communication and shared experiences in sustaining group unity, akin to entangled states in quantum systems. When these elements are lacking or weakened, decoherence sets in, leading to social fragmentation. For example, organizations undergoing rapid changes without clear communication may see a decline in collaboration, much like particles losing alignment. This insight encourages leaders to enhance communication pathways and cultivate environments promoting shared experiences, acting as a cohesive force that binds members together, preventing social fragmentation.

The concept of quantum decoherence also offers a fresh perspective on societal polarization. As individuals in a community identify more strongly with subgroups or external affiliations, the previously cohesive community begins to splinter. This mirrors a quantum system's shift from coherence to decoherence when external interactions disrupt its unified state. By exploring these parallels, sociologists can devise strategies to counteract polarization, such as fostering inclusive narratives and creating cross-group interactions that nurture a sense of collective identity, ultimately aiming to preserve social harmony.

The shift from coherence to fragmentation in social settings often involves a complex interaction of individual behaviors and external influences. Consider the role of social media, which can amplify divisive narratives and expedite fragmentation. In today's digital landscape, the rapid dissemination of information can disrupt the cohesive fabric of societal groups, akin to environmental interference causing quantum decoherence. Recognizing these influences equips individuals and communities to better navigate the challenges posed by digital interactions, emphasizing the importance of digital literacy and critical thinking to maintain social cohesion.

To effectively address social fragmentation, it is crucial to apply insights from quantum decoherence to practical scenarios. Encouraging reflection on the shared values and goals that initially united the group can help reignite a sense of common purpose. Creating opportunities for dialogue and collaboration among diverse members can strengthen the bonds holding the group together, counteracting the forces of social disintegration. By understanding the parallels between quantum phenomena and group dynamics, we can develop innovative strategies for fostering resilience and cohesion in human interactions, paving the way for more harmonious communities.

Collective Decision-Making as a Quantum Coherence Phenomenon

Collective decision-making can be understood through the analogy of quantum coherence, where groups align harmoniously toward shared objectives. This process resembles particles in a synchronized quantum state, exhibiting unified behavior that enhances overall efficiency. Within groups, coherence arises as individuals synchronize intentions and actions, propelling collective decisions in a unified direction. Research in social psychology suggests that group coherence often leads to more effective decision-making by aligning motivations and reducing internal friction. By fostering an environment of open communication and shared vision, groups can harness this coherence, improving their ability to make decisions that mirror the collective intent.

The balance between individual autonomy and group harmony is crucial in collective decision-making. While coherence implies uniformity, it doesn't erase individuality. Instead, it flourishes on the diversity of perspectives that merge into a singular, potent entity. Quantum coherence illustrates that particles maintain individual properties while contributing to a unified state. Similarly, in groups, individual contributions are essential to overall harmony. Valuing diverse voices within a group allows organizations to cultivate an atmosphere where coherence naturally emerges. This ensures decisions benefit from a rich array of ideas, fostering innovation and resilience.

Insights from quantum coherence offer valuable lessons for optimizing collective decision-making. The concept of entanglement, where particles remain interconnected over distances, serves as a metaphor for trust-based relationships within groups. Trust acts as the glue that maintains harmony, enabling seamless information exchange and mutual support. Innovative studies highlight that groups investing in trust and understanding are more likely to sustain coherence, even under external pressures. With trust, members engage in honest dialogue, leading to decisions that are coherent and sustainable.

Questions arise about what disrupts this harmony, causing the breakdown of collective decision-making. Just as quantum decoherence occurs when external forces disturb a system's equilibrium, social coherence can falter due to miscommunication, conflicting interests, or unclear objectives. Addressing these disruptors requires a proactive approach: establishing clear communication channels, defining shared goals, and creating conflict resolution mechanisms. By anticipating potential sources of disruption, groups can implement strategies to maintain alignment and reinforce decision-making processes.

For those aiming to harness quantum coherence in their groups, practical steps can provide guidance. Encourage environments that prioritize empathy and active listening, crucial for nurturing the trust and understanding that underpin harmony. Facilitate regular discussions for members to voice concerns and align on objectives, ensuring everyone is synchronized. Embrace diversity as a strength, recognizing that varied perspectives enrich decision-making. By applying these principles, groups can mirror the coherent states found in quantum systems, making decisions that are cohesive and reflective of the collective's potential.

The Transition from Coherence to Decoherence in Social Movements

In social movements, shifting from unity to fragmentation mirrors changes in the quantum realm, where systems transition from cohesive behavior to disarray. Initially, social cohesion arises when a group unites around a common goal, similar to quantum coherence where particles act in concert. This unity is propelled by a collective vision, fostering synchronized efforts and solidarity. However, increasing external pressures or internal conflicts can disrupt this harmony, akin to the quantum phenomenon of decoherence. Initial unity in social movements may fracture due to differing interests, strategies, or outside influences, leading to a loss of collective focus.

GROUP DYNAMICS AND QUANTUM COHERENCE 113

Emerging research in social psychology and quantum theory explores these transformations. Studies indicate that a strong, charismatic leader can temporarily maintain unity, much like a laser aligning particles. Yet, when leadership wanes or is challenged, decoherence becomes evident as group unity disintegrates. This dynamic is observed not only in grassroots movements but also in corporate settings and political campaigns. As movements expand, the complexity and variety of individual motivations can disrupt unity, resulting in a fragmented group struggling to achieve its original goals.

A captivating parallel between quantum mechanics and social dynamics is the concept of entanglement, where actions in one part of a system influence another, even from afar. In social movements, decisions or events in one region can ripple globally, impacting unity. This interconnectedness can either strengthen movements by spreading awareness or lead to fragmentation if messages become diluted or misinterpreted. The role of digital media in this interconnected landscape is significant, serving as both a catalyst for unity and a source of distraction that contributes to fragmentation.

The transition from unity to fragmentation in social movements isn't inherently negative; it offers opportunities for growth and adaptation. By understanding the mechanics of fragmentation, leaders and participants can develop strategies to manage disruption. Techniques like fostering open communication, ensuring transparent decision-making, and embracing diverse perspectives can help maintain unity. Emphasizing adaptable structures within movements allows for resilience in the face of change, much like how quantum systems adapt to external influences.

This dynamic prompts reflection on sustaining unity in social engagements. Consider strategies to nurture resilience amid inevitable change. How can lessons from quantum coherence enhance collaboration and innovation in diverse groups? By contemplating these questions, readers are encouraged to apply insights from quantum physics to real-world social contexts, deepening their understanding of the complex interplay between unity and fragmentation in human interactions.

By examining the intricate connections between subatomic harmony and collective dynamics, this chapter has revealed how the fundamental principles of unified physical systems echo within human communities. Recognizing social synchronization as the force that aligns collective behaviors, we uncover the unseen energies that hold societies together, akin to how particle coherence sustains the unity of atoms. This analogy extends to collective actions, which reflect subatomic harmony, while disruptions and conflicts mirror the breakdown of this coherence. These insights not only deepen our understanding of social behaviors but also highlight the potential of leveraging physics-inspired models to anticipate and shape communal actions. As we continue this exploration, consider how these concepts might revolutionize our approach to societal challenges—promoting harmony and innovation amidst complexity. With this understanding, we are poised to further investigate how quantum principles and advanced technologies can demystify human interactions, encouraging us to rethink the very essence of social life.

Chapter Seven

The Heisenberg Uncertainty Principle And Social Unpredictability

In the lively core of a city park, a group of pigeons suddenly takes flight, their movements unpredictable yet harmonized in a breathtaking display. Each bird charting its own course somehow contributes to the collective grace of the flock, offering a vivid reflection of the intricate dance of human interactions. Just as these birds navigate a realm of uncertainty, we too find ourselves in a world where the future remains a puzzle of endless possibilities. It is within this unpredictable landscape that the principles of quantum mechanics reveal profound insights about our social existence.

In this chapter, we delve into the enigmatic Uncertainty Principle, a guiding beacon for understanding the unpredictable nature of human behavior. Just as a particle's position and momentum resist simultaneous precision, human actions often slip through the grasp of exact prediction. This quantum insight encourages us to welcome life's uncertainties, celebrating the spontaneity that colors our relationships and social exchanges. By drawing parallels between the indeterminacy of particles and the variability of social behavior, we uncover why, despite our best efforts to forecast, human dynamics remain intriguingly elusive.

As we navigate these ideas, we will explore how embracing the unknown can reshape our approach to relationships and group dynamics. In the shifting dance of human connections, where predictability often yields to surprise, lies the potential for growth and innovation. By acknowledging the boundaries of certainty, we nurture resilience and creativity in our social ventures. Throughout this journey, the principles of quantum mechanics illuminate our path, challenging us to reconsider the nature of predictability and guiding us toward a richer understanding of the social world we inhabit.

The Uncertainty Principle: Limits of Predictability in Physics

At the crossroads of theory and practice lies the fascinating interplay of uncertainty, a concept that defies our quest for precision and control. In the realm of quantum mechanics, unpredictability is not an anomaly but a defining feature, showcasing a universe where observation itself alters reality. This world, governed by the principles of the uncertainty principle, starkly contrasts with the predictability championed by classical physics. Here, the challenge is not in perfecting tools or formulas but in embracing the intrinsic boundaries of knowledge. As we journey through the quantum lens, it becomes clear that the erratic behavior of particles is not a flaw, but a fundamental truth. This insight encourages us to rethink our understanding of predictability, especially within human interactions.

Consider human relationships akin to a complex quantum system, where every action is shaped by myriad influences, many unseen. Just as measuring one element in a quantum system can disturb another, the subtleties of human dynamics defy complete foresight. Social bonds and communities are ever-shifting mosaics, formed by a blend of personal decisions, external factors, and the subtle effects of observation. As we delve deeper, the connections between quantum mechanics and human behavior become apparent, highlighting why the unpredictability in the atomic realm is mirrored in human actions. This exploration aims to shed light on the complexities of social unpredictability, offering perspectives on navigating the uncertainties within our connections and communities.

Quantum Mechanics and the Nature of Uncertainty

Quantum mechanics transforms our perception of uncertainty, challenging the predictable nature of classical physics. Central to this is the uncertainty principle, which asserts that pairs of physical properties, such as position and momentum, cannot be measured with absolute precision at the same time. This limitation is not due to technological shortcomings; rather, it is a fundamental characteristic of quantum systems. When we achieve precise measurement of one property, the other becomes uncertain. This principle invites a reconsideration of reality, emphasizing the intricate relationship between measurement and knowledge.

The uncertainty principle introduces a realm where precision has boundaries, urging a reevaluation of predictability. Consider electrons within an atom: they do not adhere to fixed orbits but exist within probabilistic clouds, where exact position and momentum remain elusive. This probabilistic nature of quantum entities challenges the classical idea of certainty, suggesting that the universe operates on probabilities rather than definites. Grasping this concept enriches our understanding of quantum mechanics and offers a metaphor for the unpredictability present in social interactions.

Comparing quantum systems to classical ones reveals a new dimension of unpredictability. Classical physics, underpinned by Newtonian mechanics, assumes that with ample information, future states can be accurately forecasted. However, quantum mechanics introduces an inherent ambiguity, where particles appear in multiple states until observed. This duality challenges traditional scientific methods, prompting a shift in how we perceive the limits of predictability. Through this lens, parallels with human behavior become apparent, where actions often defy straightforward predictive models.

The observer effect adds complexity to our understanding of uncertainty within quantum systems. In the quantum world, observation alters the state of particles, affecting the system being observed. This interaction between observer and observed raises compelling questions about the essence of reality and knowledge. Extending this idea to the social realm, human interactions are similarly influenced by the presence of an observer, whether it be a leader in a group or societal norms shaping behavior. Recognizing this parallel fosters exploration into how awareness and perception can modify social dynamics, much like the observer effect influences quantum states.

By embracing quantum uncertainty, we gain valuable insights into managing unpredictability in both physics and social interactions. This perspective encourages us to move beyond rigid expectations and adopt a mindset that values complexity and ambiguity. Practically, acknowledging the limits of predictability can lead to more resilient strategies in decision-making, whether in scientific pursuits or navigating human relationships. As we explore these intersections, we deepen our understanding of the universe and ourselves, fostering a shift towards more nuanced interpretations of behavior and interaction.

Measuring Particles and the Impact on Predictability

The core of quantum mechanics presents a fascinating challenge: measurement, which deeply influences our grasp of predictability. At the quantum level, when scientists examine particles, they face a fundamental constraint—precisely determining a particle's position results in less certainty about its momentum, and vice versa. This concept, known as the Uncertainty Principle, underscores the inherent unpredictability of quantum systems. Unlike classical physics, where positions and velocities can be calculated with accuracy, quantum mechanics embraces the probabilistic essence of existence. This framework alters our approach to forecasting and certainty, offering profound insights into the fabric of reality.

Consider the analogy of a dart in flight. In classical physics, every detail of its movement is trackable, allowing predictions about its path with confidence. Yet in quantum mechanics, the dart seems to occupy all positions simultaneously. Measuring one property—like its velocity—makes its precise location elusive. This duality adds complexity, challenging conventional wisdom by suggesting that measurement alters the observed system. The implications reach beyond physics, influencing our understanding of intricate systems like human societies.

Recent breakthroughs in quantum research reveal the subtle interaction between measurement and predictability. Techniques such as weak measurement enable physicists to gather partial information without fully collapsing a quantum state, offering a fresh perspective on uncertainty. This approach parallels social sciences, where observing partial behaviors can provide insights without needing complete data. By accepting partial predictability, researchers uncover patterns and trends that might otherwise remain hidden. This mirrors how AI and machine learning models predict social behaviors, identifying trends within vast data sets without claiming absolute certainty.

Reflecting on the implications of quantum measurement in social contexts offers a thought-provoking perspective on human behavior. Imagine a group of people, where each individual's actions and interactions are in constant flux. While general patterns within the group might be predictable, the exact behavior of any single person remains elusive. This mirrors quantum measurement's uncertainty, where the collective behavior of particles can be estimated, yet individual particles defy precise prediction. Such scenarios invite deeper reflection on the nature of knowledge and the limits of predictability in both quantum and social systems.

Armed with this understanding, one can explore practical insights for navigating uncertainty. Embracing the unpredictability inherent in both quantum systems and human interactions can lead to more adaptable strategies across various fields—be it in business, diplomacy, or personal life. Instead of striving for absolute certainty, recognizing the value of flexibility and resilience can enhance decision-making processes. By appreciating the subtleties of measurement and its impact on systems, we cultivate a deeper appreciation for the complexities of both the quantum realm and the intricacies of human society.

Uncertainty in Quantum Systems Versus Classical Systems

In the domain of quantum mechanics, uncertainty is a fundamental feature, setting it apart from the predictability inherent in classical physics. At the heart of this are principles like Heisenberg's Uncertainty Principle, which asserts that certain pairs of physical properties, such as position and momentum, cannot be pinpointed with absolute precision simultaneously. This intrinsic uncertainty contrasts sharply with the deterministic nature of classical systems, where accurate measurements and forecasts are feasible. The very act of observing a quantum system disturbs it, introducing a core element of unpredictability. This notion challenges the Newtonian paradigm, which envisions a universe unfolding with precision, suggesting instead a reality governed by probabilities.

To comprehend this uncertainty, consider the behavior of electrons within an atom. In classical physics, electrons might be envisioned as orbiting the nucleus akin to planets around a sun, their paths precisely calculable. However, within a quantum framework, electrons exist in a probabilistic cloud, their exact position and momentum indeterminate until observed. This probabilistic nature is not a flaw but an essential characteristic of quantum systems. The dual nature of particles behaving as both waves and particles further complicates predictability, as demonstrated by the renowned double-slit experiment. When electrons pass through two slits, they form an interference pattern, behaving like waves, until they are observed, at which point they appear as particles. This phenomenon highlights the constraints imposed by the uncertainty principle and underscores the contrast with classical predictability.

The observer effect is crucial in this quantum landscape, showing how observation impacts the system being measured. This is not merely a technical limitation but a profound implication of quantum mechanics, where the act of observation alters the phenomenon itself. In classical physics, observation is passive, revealing what already exists. In the quantum realm, however, observation actively shapes reality, challenging conventional notions of objectivity and predictability. This interaction between observer and system prompts reflection on the nature of knowledge and the limits of human understanding.

Recently, researchers have harnessed quantum uncertainty to explore groundbreaking areas like quantum cryptography and quantum computing. By leveraging principles such as superposition and entanglement, these technologies offer unparalleled security and computational power. Quantum cryptography, for instance, employs the uncertainty principle to create unbreakable codes, as any eavesdropping attempt alters the quantum state, exposing the intrusion. Similarly, quantum computers exploit the probabilistic nature of qubits to perform calculations far beyond the capabilities of classical computers. These advances highlight the potential of using uncertainty as a tool rather than a barrier, offering transformative possibilities across various fields.

Exploring the unpredictable terrain of quantum systems provides valuable insights into the complexities of social dynamics. Just as quantum mechanics challenges the certainty of classical physics, human behavior often defies simplistic predictions. Recognizing the parallels between these domains invites a reevaluation of our expectations of predictability in both the physical and social worlds. By embracing the uncertainty inherent in quantum systems, we gain not only a deeper appreciation for the mysteries of the universe but also a richer understanding of the intricacies of human interactions. This perspective encourages a shift from seeking absolute certainty to fostering adaptability and resilience in the face of unpredictability.

The Role of Observer Effect in Quantum Predictability

In the fascinating world of quantum mechanics, the observer effect illustrates how observation can fundamentally change a particle's behavior. This phenomenon highlights the intricate relationship between measurement and uncertainty, demonstrating that observing a particle's position or momentum limits our ability to predict its future state precisely. This insight challenges our traditional notions of predictability, suggesting that the universe actively participates in reality's unfolding rather than passively awaiting discovery. This perspective not only reshapes our understanding of the physical world but also draws intriguing parallels to the unpredictability of human behavior.

Consider how the observer effect might play out in social contexts, where interactions are often swayed by the presence and perception of others. Similar to how observing a particle alters its state, the awareness of being watched can change individual behavior, affecting decisions and social dynamics. This concept highlights the inherent unpredictability within social systems, as knowledge of observation can lead individuals to act differently than they would privately. Understanding this parallel encourages us to rethink how social environments, much like the quantum realm, are shaped by observation and interaction, where outcomes depend on the ongoing influence of observation.

Recent research in quantum mechanics has further nuanced our understanding of the observer effect, showing that the impact of observation on particles varies with the measurement's nature and context. These discoveries suggest that predictability in both quantum and social systems depends on specific parameters and contexts of observation. In social settings, this suggests that human behavior's predictability is influenced by factors like cultural norms, social expectations, and awareness of observation. Recognizing these variables allows for a deeper appreciation of social predictability and the challenges of forecasting human actions with certainty.

As we explore the observer effect, we are prompted to consider how observation influences individual actions and shapes group dynamics. The presence of observers, whether real or perceived, can lead to phenomena like groupthink or conformity, where individuals align with perceived group norms. This dynamic mirrors the quantum reality, where particles exhibit collective behavior in response to observation, indicating that the observer effect significantly influences both micro-level interactions and macro-level social trends. Such insights invite reflection on the power of observation and perception in shaping societal structures and the interplay between individual choices and collective influence.

As we navigate uncertainties in both the quantum world and social interactions, the observer effect encourages a nuanced view of predictability and control. While definitive answers and absolute predictability may be tempting, both quantum mechanics and social sciences remind us that uncertainty is an inherent and valuable aspect of complex systems. Embracing this uncertainty opens new possibilities for understanding and influencing social dynamics. Recognizing that observation is an active engagement with the world's complexities empowers us to approach social interactions with greater empathy and adaptability, acknowledging the fluid nature of human behavior in an ever-evolving landscape.

Social Uncertainty: Why Human Behavior Can Never Be Fully Predicted

Imagine a world where every choice and action unfolds with perfect predictability. While this might initially seem captivating, human behavior is a far richer and more intricate tapestry. Much like the limits highlighted by the uncertainty principle in quantum physics, which states the impossibility of precisely knowing both the position and momentum of a particle, the unpredictability of social interactions is a fundamental trait rather than a flaw. This very unpredictability weaves the vibrant fabric of human experience, characterized by spontaneity, creativity, and diversity. Just as quantum particles resist deterministic predictions, human actions are influenced by countless factors that interact in surprising ways.

This inherent uncertainty in human behavior is shaped by a blend of elements that challenge simple explanations. Free will is a key factor, allowing individuals to make choices that sometimes defy logic. Cultural and social norms add another layer of complexity, subtly influencing both individual and group behaviors. Emotions further complicate decisions, often overshadowing cold facts. The dynamic interplay between personal choices and group dynamics creates a constantly evolving landscape, where individual actions resonate through social networks in unexpected ways. By delving into these elements, one gains a deeper appreciation for the complexity and beauty of human interactions, encouraging readers to embrace uncertainty as an essential aspect of the social experience.

The Role of Free Will in Human Decision-Making

The idea of free will in human decision-making sits at a fascinating crossroads of philosophy, neuroscience, and social science, presenting a complex mosaic of insights into the unpredictable nature of behavior. When considered alongside the Uncertainty Principle, which highlights inherent unpredictability, free will adds another layer of intricacy to our comprehension of human actions. As individuals maneuver through their daily routines, the decisions they make are shaped by numerous influences, yet free will suggests a spontaneity that evades complete anticipation. This autonomy resembles the unpredictable motion of particles that, despite existing within a framework of laws and forces, demonstrate behaviors that cannot be entirely forecasted. Just as quantum mechanics accepts uncertainty as essential, we must also acknowledge the unpredictability inherent in human decision-making.

Recent breakthroughs in neuroscience have started to reveal the neural foundations of free will, offering captivating insights into how the brain forms choices. Although the brain operates within a network of neural pathways and chemical signals, the emergence of consciousness and subjective experience implies a level of agency that goes beyond mere biological processes. Research using fMRI technology has shown instances where decisions occur before individuals become consciously aware of them, challenging traditional ideas of free will while underscoring the complexity of predicting human behavior. This burgeoning research indicates that while certain patterns can be discerned, unpredictability remains a fundamental aspect of human nature, akin to the probabilistic tendencies seen in quantum systems.

The role of free will in decision-making becomes even more intricate due to the dynamic interplay between personal autonomy and external influences. Social norms, cultural expectations, and personal experiences collectively create the backdrop for decision-making, yet free will empowers individuals to navigate and occasionally resist these pressures. This interaction can be compared to the dual nature of particles that exhibit both wave-like and particle-like properties, transcending simple definitions. The exercise of free will is not isolated but is in constant interaction with the social environment, creating a dance of choice and influence that defies straightforward prediction. The challenge lies in comprehending how these elements come together to guide decisions in ways that are neither entirely random nor wholly deterministic.

When considering the implications of free will, it is crucial to explore how this concept affects group dynamics and collective decision-making. When individuals assert free will within a group, their actions can influence the direction of collective outcomes, introducing an unpredictability similar to the observer effect in quantum physics, where observation alters the state of the system. In social contexts, each person's decisions affect and are affected by others, creating a feedback loop that amplifies uncertainty. This complexity prompts a reevaluation of traditional predictive models, encouraging a more nuanced approach that considers both individual agency and the interconnectedness of group behavior.

To navigate the inherent uncertainty in human decision-making, it is vital to cultivate an awareness of the factors influencing our choices and remain open to the possibility of change. Embracing the uncertainty of free will can lead to more adaptive strategies in personal and professional settings, recognizing that while we may not predict every outcome, we can prepare for a range of possibilities. By fostering environments that encourage open dialogue and flexibility, individuals and groups can better harness the creative potential of unpredictability, transforming it from a source of anxiety into a catalyst for innovation and growth. Recognizing the parallels between quantum uncertainty and human free will allows us to appreciate the intricate interplay of choice and chance that defines our social existence.

The Influence of Cultural and Social Norms on Behavior

Cultural and social norms create the unseen framework that shapes human behavior, guiding actions and decisions in ways that are often profound yet subconscious. These norms serve as guiding principles, impacting individuals through collective beliefs and societal expectations. They are crucial in maintaining social order but also introduce unpredictability. Norms can change over time, sometimes swiftly, as societies evolve, and what is deemed acceptable in one context may be entirely different in another. This variability adds complexity to predicting human behavior, as norms are dynamic and continually renegotiated within communities. Recent sociological research indicates that individuals often navigate a complex landscape of conflicting norms, balancing personal desires with societal expectations, thus adding layers of intricacy to behavior prediction.

Consider how cultural norms influence decision-making. In collectivist societies, decisions often prioritize group well-being, emphasizing harmony and consensus. Conversely, individualistic cultures might focus on personal autonomy and self-expression. These cultural perspectives can lead to vastly different behaviors in similar situations. This dichotomy is further complicated by globalization, as individuals are increasingly affected by multiple cultures simultaneously. The interplay between cultural forces and personal values generates a tapestry of motivations that defy simple prediction models. Advanced research in cross-cultural psychology highlights the importance of understanding these nuances, revealing the intricate ways cultural norms dictate behavioral variability across populations.

While similar to cultural norms, social norms operate more immediately, often within specific groups or communities. They dictate expected behaviors in various settings, such as workplaces, schools, or social gatherings. Social norms wield power by enforcing conformity or encouraging deviation. For example, peer pressure illustrates how social norms can lead individuals to act against their better judgment, highlighting the tension between individual agency and collective influence. Recent findings in social network analysis show that the strength and reach of social networks can significantly amplify or mitigate the impact of these norms, revealing another layer of complexity in predicting behavior. Understanding these network mechanics is crucial for developing more accurate models of human behavior.

In the realm of social uncertainty, the emotional landscape of individuals is vital. Emotions intertwine deeply with cultural and social norms, serving as both drivers and responses to norm adherence or violation. Emotional reactions can reinforce norms, as seen when guilt or shame arises from norm violations, or they can challenge them, as when righteous anger fuels social change. The emotional complexity inherent in human interactions adds an unpredictable element to any behavioral forecast. Research in affective science is beginning to unpack these emotional dynamics, providing valuable insights into how emotions interact with norms to influence behavior in both predictable and surprising ways.

To navigate the unpredictable terrain shaped by cultural and social norms, individuals and organizations can employ strategies that embrace flexibility and adaptability. Recognizing the fluid nature of norms and the diverse motivations that drive behavior can foster a nuanced understanding of social interactions. This involves actively seeking diverse perspectives, being open to cultural learning, and fostering environments where dialogue about norms is encouraged. Such approaches not only enhance the ability to anticipate behavior but also promote more harmonious and inclusive societies. These strategies, grounded in the latest research and insights, offer practical ways to manage the inherent unpredictability of human behavior in a world defined by ever-shifting norms.

Emotional Complexity and Its Impact on Predictability

The intricate nature of human emotions presents a formidable challenge to predicting behavior, weaving a complex tapestry that resists simple interpretation. Emotions are inherently layered and deeply entwined with our thoughts and life experiences, affecting decisions in often unexpected ways. Recent advances in neuroscience and psychology highlight the significant role emotions play in cognitive processes, showing they are not just reactions but integral to decision-making. This complexity is evident when exploring the diverse factors influencing emotional responses, from personal backgrounds and histories to cultural contexts and societal pressures. These elements interact dynamically, meaning the same stimulus might provoke varied responses in different individuals, adding layers of uncertainty to human interactions.

This unpredictability in emotions can be compared to the superposition principle in quantum mechanics, where various potential states coexist until observed. In social settings, emotions can simultaneously embody conflicting states—such as affection and bitterness, happiness and grief—each unexpectedly influencing behavior. This dual nature complicates efforts to foresee human actions, as pinpointing an individual's exact emotional state at any moment can be elusive, even to the person themselves. The burgeoning field of affective computing seeks to unravel this complexity using AI to analyze emotional signals, yet the subtlety of emotions often escapes even the most advanced algorithms.

Emotional complexity goes beyond individual responses, permeating social groups and communities. Collective emotions, which emerge when people share emotional experiences, can heighten unpredictability in group dynamics. Emotional contagion, where emotions rapidly spread within a group, powerfully illustrates this. Such phenomena can lead to unexpected shifts in group behavior, as seen in social movements or market trends driven by collective sentiment rather than logical analysis. Understanding these dynamics requires a comprehensive approach that considers the intricate interdependencies within social networks, challenging researchers to develop innovative models that capture the fluid and emergent nature of group emotions.

Navigating the maze of emotional complexity calls for a nuanced approach, one that recognizes inherent uncertainty while seeking paths to clarity. Individuals and organizations can benefit from developing emotional intelligence, the ability to recognize, understand, and manage emotions in oneself and others. By fostering emotional awareness and empathy, people can better anticipate emotional responses and adapt their interactions accordingly, reducing friction and promoting harmony. This approach aligns with principles akin to quantum thinking, embracing ambiguity and uncertainty as opportunities for deeper connection and understanding.

Reflecting on the unpredictable nature of emotions invites us to reconsider the frameworks through which we view human behavior. While complete predictability may remain elusive, embracing the richness of emotional complexity can inspire new ways of thinking and interacting. By acknowledging the role of emotions in shaping our social world, we open the door to more compassionate and adaptive forms of engagement, where the focus shifts from controlling outcomes to understanding the intricate dance of human emotions. This perspective not only enriches our personal and professional lives but also offers a profound insight into the very essence of what it means to be human.

The Dynamic Interplay Between Individual Choices and Group Dynamics

In the complex world of social dynamics, the interaction between individual choices and collective behavior creates a fascinating mosaic marked by unpredictability. Each person's decisions are shaped by various influences, weaving a network of interactions that can alter group behavior. This phenomenon, reminiscent of the interactions within a quantum system, presents a captivating subject for examination. While individuals possess free will, their choices are inevitably influenced by the broader social environment. As people interact, their decisions send ripples through the group, resulting in emergent patterns that can be surprising. This interconnectedness highlights the challenge of predicting social behavior with certainty, as the collective outcome transcends the sum of individual actions, forming a complex web of interdependencies.

Recent research in social physics emphasizes how individual actions can resonate within a group, contributing to phenomena like the spread of trends or the swift adoption of new norms. Consider viral social media challenges, where one person's action, when in tune with the cultural zeitgeist, can cascade through networks, impacting many others. This amplification mirrors quantum entanglement, where particles remain linked across distances, suggesting that human interactions can extend beyond immediate social boundaries. The unpredictability of these social cascades challenges traditional models of behavior prediction, prompting a reevaluation of how influence and decision-making are understood in social contexts. By acknowledging the quantum-like nature of social dynamics, new models can be developed that account for the intricate nuances of these interactions, offering deeper insights into the mechanisms driving collective behavior.

The relationship between individual agency and social influence can be further explored through game theory and network analysis. These fields provide frameworks for understanding decision-making within social networks, considering both personal incentives and potential reactions from others. Game theory shows that individuals often make choices based on not only personal preferences but also anticipated responses from their peers. This strategic decision-making process highlights feedback loops inherent in social interactions, where individual actions are continually adjusted based on the evolving group dynamics. Network analysis complements this by illustrating the connections and pathways through which influence travels, creating a visual representation of potential change within a community. Together, these approaches underscore the importance of considering both individual and collective factors when understanding social behavior.

Incorporating insights from quantum mechanics, researchers have started developing innovative models that embrace the uncertainty inherent in social interactions. These models move beyond linear predictions, embracing the probabilistic nature of social outcomes. By integrating principles, such as superposition, where multiple outcomes coexist until a decision is made, these frameworks offer a more nuanced understanding of how social dynamics unfold. This approach not only aligns with cutting-edge research in social science but also provides practical tools for navigating the complexities of human interaction. For instance, organizations can apply these models to anticipate shifts in public opinion or the impact of policy changes, enhancing their ability to adapt to an ever-changing social landscape.

Engaging with the dynamic interplay of individual and group dynamics encourages reflection on how we navigate our social environments. By viewing the unpredictability of social interactions as an opportunity rather than a limitation, individuals and organizations can cultivate adaptability and resilience. Thought-provoking questions arise: How can we harness this unpredictability to foster innovation? What strategies can be used to manage the influence of group dynamics on personal decision-making? By embracing a quantum perspective, readers are encouraged to approach their social experiences with curiosity and openness, recognizing that the seemingly chaotic nature of human interactions holds potential for creativity and growth. As we deepen our understanding of these dynamics, we unlock new possibilities for collaboration and understanding in an increasingly interconnected world.

Navigating Uncertainty in Relationships and Social Groups

In the intricate ballet of human relationships, uncertainty threads its way through our lives, echoing the unpredictable dance of tiny particles. To truly understand its impact on our interactions and communities, we must recognize uncertainty as a vital element, not just a challenge to overcome. Similar to how the Uncertainty Principle in physics highlights the limits of precision, human connections are characterized by a degree of ambiguity, a realm where multiple possibilities coexist and outcomes can change. This inherent unpredictability invigorates our bonds, offering opportunities for growth, transformation, and exploration. In this light, ambiguity becomes a powerful driver for creativity and adaptation, encouraging us to discover fresh ways of engaging with one another.

Navigating this uncertainty demands a deep understanding of perception and its influence on group behavior. Much like the observer effect in quantum physics, perception shapes how we interpret and react to social signals, impacting collective dynamics. As we interact within our social circles, we face both challenges and opportunities in managing these diverse perceptions, requiring strategies that build resilience and flexibility. Tools informed by AI and insights from quantum theories present intriguing possibilities for anticipating changes in our connections. While not perfect, these models offer a framework to explore the complex interplay of relationships, allowing us to appreciate the beauty of uncertainty and the limitless potential it holds for individual and communal development.

Embracing Ambiguity in Personal Connections

In a world where human connections often resist clear-cut explanations, embracing uncertainty becomes both a challenge and an opportunity. Much like the unpredictability observed in quantum mechanics, human interactions exhibit a similar complexity. This inherent variability, rather than obstructing relationships, enriches them with depth and nuance. It encourages individuals to practice patience and openness, understanding that not every interaction or feeling can be easily categorized. By embracing this uncertainty, people nurture more authentic connections, allowing relationships to evolve naturally without the burden of rigid expectations.

An intriguing facet of uncertainty in relationships is the power of perception. Similar to quantum phenomena, where observation can influence a particle's state, our perceptions within social settings can alter interaction dynamics. Recent social psychology studies indicate that our beliefs and expectations about others can significantly shape their behavior. This highlights the importance of mindfulness in communication, as the way we perceive others can steer the course of the relationship. Fostering an awareness of this perceptual influence can lead to more empathetic and adaptable interactions, where individuals remain open to the complexity of human behavior.

Navigating uncertainty also involves tackling the unpredictability found within social groups. The intricate dynamics of groups often resemble the interconnection of quantum particles, where one person's actions can impact the entire system. Embracing uncertainty here requires balancing personal agency with collective influence. Advanced research into group behavior shows that flexibility in roles and expectations can enhance group cohesion and innovation. By welcoming diverse perspectives and unexpected outcomes, groups can use uncertainty as a catalyst for creative problem-solving and growth.

In practical terms, handling uncertainty in relationships involves adopting strategies that promote resilience and adaptability. Techniques like active listening and open communication can create environments where uncertainty is not just tolerated but valued. Encouraging a culture of curiosity—where questions are welcomed and assumptions are challenged—can transform potential confusion into opportunities for learning and development. By integrating these strategies into daily interactions, individuals can forge stronger, more resilient networks that thrive amidst unpredictability.

Ultimately, embracing uncertainty in personal connections paves the way for richer, more fulfilling relationships. It invites individuals to transcend conventional narratives, exploring the vast landscape of human emotions and interactions with wonder and openness. By recognizing that uncertainty is intrinsic to the human experience, people can cultivate deeper connections that adapt to life's unpredictable nature. This approach not only enhances personal relationships but also enriches the broader social fabric that binds communities together.

The Role of Perception in Group Dynamics

Perception significantly influences group dynamics, acting as a filter through which individuals understand the actions and motivations of others. In any group setting, the perceptions of its members can greatly impact how they interact and make decisions. These perceptions are dynamic, evolving through social interactions and influenced by both personal biases and collective experiences. Just as quantum physics suggests that observation can change a particle's state, social perception can also alter group dynamics. When individuals expect certain behaviors based on previous interactions, they may inadvertently encourage those behaviors, creating a feedback loop that solidifies their initial perceptions. This highlights the substantial effect that shared mental models and expectations have on the cohesion and functionality of a group.

Understanding perception's role in groups reveals the complexities of human connections, where subjective views can lead to outcomes ranging from cooperation to conflict. Research in social psychology indicates that preconceived ideas about group members can distort interpretations of their actions, possibly leading to misunderstandings or disputes. In professional environments, if a team member is seen as overly critical, their constructive feedback might be dismissed or viewed suspiciously, hindering effective collaboration. Recognizing this dynamic provides a way to reduce potential tensions by promoting open dialogue and challenging assumptions. Encouraging environments where perceptions can be openly discussed helps groups better navigate human interactions.

In the digital era, perception's impact on group dynamics extends beyond face-to-face interactions. Online platforms magnify the role of perception, as individuals often form opinions based on limited information, like profile images or brief messages. This can lead to quick, often incorrect, judgments about group members, affecting communication and subgroup formation. Virtual settings present unique challenges and opportunities for managing perception within group dynamics. Technology can be leveraged to provide more context and deepen understanding, countering the shallow judgments that digital interactions often encourage. By using tools that foster comprehensive exchanges, groups can align perceptions more closely with reality, promoting cohesive and productive interactions.

Exploring perception's influence on group dynamics also requires consideration of cultural and societal factors. Diverse cultural backgrounds can shape how individuals see group norms and behaviors, affecting their engagement with others. In multicultural teams, the interplay of different cultural perceptions can either enhance the group's dynamic or cause tension if not properly managed. Valuing and recognizing these diverse perspectives can boost creativity and innovation within a group. Organizations that prioritize cultural awareness and sensitivity can turn potential challenges into strengths, using varied viewpoints to encourage more inclusive and effective collaboration. By embracing these differences, groups can build more resilient and adaptable structures.

Practical strategies for managing perception in group dynamics include creating an environment where openness and empathy are prioritized. Encouraging members to express their perceptions and the reasoning behind them can uncover hidden biases and promote mutual understanding. Presenting scenarios where group members interpret ambiguous situations can reveal diverse perspectives, fostering empathy and reducing judgmental attitudes. Regular feedback sessions and reflective practices can help adjust perceptions, ensuring they are more aligned with reality. By fostering trust and transparency, groups can effectively navigate the complexities of perception, enhancing their ability to adapt and thrive in a continually evolving social landscape.

Strategies for Managing Uncertainty in Social Networks

In our interconnected world, social networks reflect the complexity of quantum systems, where the unpredictability of human relationships demands innovative approaches that blend scientific insights with human intuition. Embracing the inherent uncertainty of social interactions, much like the nature of quantum mechanics, allows for a mindset that values adaptability. This approach enables individuals to gracefully navigate the dynamic nature of interpersonal connections.

Perception plays a crucial role in these networks. Similar to the observer effect in quantum theory, the way people perceive and interpret social cues can significantly shape group dynamics. This interaction between perception and behavior highlights the need for a balance between subjective understanding and objective analysis. By utilizing advanced AI and data analytics, individuals and organizations can gain deeper insights into these dynamics, making well-informed decisions that consider both visible and hidden forces.

To tackle uncertainty in social networks effectively, it's crucial to develop strategies that integrate intuition with empirical evidence. One effective strategy is the use of predictive models powered by machine learning to analyze patterns in social interactions. These models provide valuable foresight into possible outcomes, helping individuals anticipate changes and adjust their approaches. Leveraging AI in decision-making enables better navigation through social complexities, reducing anxiety and fostering harmonious relationships.

The balance between uncertainty and stability in social networks prompts a reevaluation of traditional relationship paradigms. In today's interconnected world, thriving amidst uncertainty is a vital skill. Promoting open communication and cultivating a culture of trust can mitigate unpredictability, resulting in more cohesive networks. Developing emotional intelligence and empathy further enhances one's capacity to manage uncertainty, facilitating understanding and cooperation among diverse individuals.

As society evolves, the challenge of managing uncertainty in social networks remains constant. By adopting quantum-inspired insights and novel methodologies, individuals can turn this challenge into an opportunity for growth and innovation. Intriguing questions arise: How can we create environments that view uncertainty as a catalyst for creativity? What role does technology play in bridging gaps and fostering deeper connections? By reflecting on these questions, readers are encouraged to apply these concepts in their own lives, redefining their approach to social networks in a practical and enlightened way.

Predictive Models of Relationship Evolution

Predictive models for understanding the progression of relationships represent a captivating intersection where science meets the intricacies of human connections. These models strive to forecast how relationships may evolve by synthesizing insights from disciplines such as psychology, sociology, and advanced AI inspired by quantum theory. Unlike traditional models that depend on static information, modern approaches leverage dynamic systems that learn from changing social inputs. By factoring in emotional states, social contexts, and external influences, these models aim to encapsulate the fluid essence of human connections, offering glimpses into potential future interactions.

In developing these predictive frameworks, researchers draw parallels between the uncertainty inherent in quantum physics and the unpredictability of human behavior. Just as quantum mechanics highlights limitations in predicting both position and momentum, social scientists acknowledge the inherent unpredictability of human actions. These models, however, focus on informed possibilities rather than certainties, providing probabilistic insights into likely outcomes. This approach mirrors the complex nature of human relationships, where numerous variables intertwine to create a web of potential scenarios.

Advanced machine learning techniques, particularly those inspired by quantum computing, hold promise for boosting predictive accuracy. These methods can process massive datasets and uncover patterns that may escape conventional techniques. For example, algorithms inspired by quantum principles can explore multiple relationship scenarios simultaneously, akin to the superposition of states in quantum mechanics. This capability enables the development of predictive models that are not only sophisticated but also aligned with the nuanced nature of human interactions.

These predictive models can be practical tools for individuals and communities navigating social complexities. By understanding potential relational trajectories, people can make more informed decisions about their interactions, fostering healthier and more resilient connections. In a professional context, predictive insights might assist in managing team dynamics by anticipating potential conflicts or synergies, enabling leaders to address issues proactively. In personal relationships, these models can guide individuals in nurturing connections that align with their long-term goals and values, ultimately enhancing satisfaction and stability.

However, it is crucial to approach these predictive models with caution and ethical consideration. While they offer intriguing insights, they should not be seen as deterministic or infallible. Human relationships are complex and influenced by free will, cultural nuances, and unforeseen events. Therefore, these models should serve as guides rather than prescriptions, inviting individuals to remain open to the myriad possibilities that define human experience. By blending scientific precision with an appreciation for human complexity, predictive models of relationship evolution can illuminate paths forward while respecting the rich tapestry of human life.

In the complex realm of human interactions, the concept of uncertainty, much like the principles found in quantum mechanics, offers a fascinating lens through which we can view our relationships. Just as physics suggests limits to precisely determining a particle's properties, social dynamics exhibit a similar unpredictability that influences our connections and group behaviors. Throughout this chapter, we've explored how this uncertainty shapes our interactions, highlighting the notion that while we may strive for understanding and prediction, absolute certainty is an elusive goal. By accepting the inherent variability in our social world, we are encouraged to approach our connections with curiosity and openness, fostering genuine and meaningful relationships. Recognizing these uncertainties allows us to build a more flexible and resilient social fabric. As we continue to explore the parallels between quantum principles and societal changes, we're invited to rethink and redefine the possibilities of human interaction, opening new avenues for understanding our evolving social landscape.

Chapter Eight

Quantum Probability And Social Decision Making

On a crisp morning in 1920s Copenhagen, Niels Bohr, a trailblazer in the world of particle physics, stood before his students and asked a question that would reshape our grasp of the universe: "How do we discuss what we cannot directly see?" Though inspired by the enigmatic world of subatomic particles, this question echoes the intricacies of human choices. Just as particles remain in a myriad of states until observed, our decisions and social behaviors exist in a web of possibilities, swayed by countless unseen influences. This chapter delves into the intriguing parallels between the probabilities of quantum mechanics and the seemingly erratic nature of societal trends.

Picture every choice you make as a roll of dice in the quantum realm, where outcomes are not merely luck but a complex interplay of possibilities. In quantum mechanics, uncertainty is not a mere measurement flaw but a fundamental aspect of reality. Likewise, in society, unpredictability shapes our interactions and collective actions. With its expanding ability to sift through vast data, artificial intelligence serves as a contemporary guide, helping us decode these probabilistic patterns. By exploring how AI forecasts social phenomena, we gain a deeper understanding of the dynamics that govern our behaviors, akin to the probabilities guiding quantum particles.

As we journey through this chapter, you'll see how quantum ideas illuminate the tapestry of social decision-making. From the ever-changing popularity of cultural trends to the shifting waves of public sentiment, these occurrences resemble the probabilistic dance of particles, ever-changing yet mysteriously unified. Embracing this perspective opens us to a richer comprehension of human interactions, challenging the deterministic views that have long dominated social sciences. With this fresh insight, the intricate interplay between chance and choice becomes not only clearer but also profoundly linked with the universe's fundamental principles.

Probability in Quantum Mechanics: The Role of Uncertainty

Imagine a fascinating frontier where the mysteries of quantum physics intersect with the complexities of human nature. Here, the tiny dance of subatomic particles mirrors the unpredictable tapestry of our social lives. Just as uncertainty governs the quantum realm, our interactions often hinge on possibilities rather than certainties. This chapter invites you on a journey to explore how principles like the Heisenberg Uncertainty Principle, a cornerstone of particle physics, resonate in the fluid dynamics of social landscapes. Much like quantum particles, our choices and relationships reflect a world of endless potential, guided by probabilities.

As we delve deeper into this intriguing convergence, picture a universe where waves of chance not only dictate a particle's path but also influence the collective decisions within communities. Social behaviors, akin to quantum systems, are ever-changing, driven by unseen forces that shape our actions. Random quantum fluctuations can illuminate the rise and fall of trends, molded by myriad factors eluding precise foresight. By weaving together artificial intelligence with quantum uncertainty, we unlock a powerful method for enhancing social predictions. AI's capacity to process extensive datasets and identify trends transforms these insights into practical tools, offering fresh perspectives to navigate the intricacies of our societal interactions. This exploration naturally leads us to examine how these theories influence group dynamics and societal outcomes, highlighting AI's potential to forecast future trends.

Exploring the Heisenberg Uncertainty Principle in Social Contexts

The Heisenberg Uncertainty Principle, a fundamental idea in quantum mechanics, suggests that certain pairs of physical properties, such as position and momentum, cannot be precisely measured at the same time. This principle can be metaphorically applied to social situations, providing an intriguing perspective on human decision-making. In social settings, the uncertainty principle highlights the unpredictability of human interactions, where the mere act of observation can influence the outcome. Imagine a scenario where an individual's decision in a group is swayed by an observer's presence, similar to how measuring a particle's position affects its momentum. This dynamic is more than just a limitation; it's an opportunity to explore the fluid nature of social behavior, acknowledging that absolute certainty is often unattainable.

Looking deeper, the concept of uncertainty in social contexts suggests that human decisions are rarely made in isolation. Instead, they are shaped by numerous external and internal factors, including cultural norms, personal biases, and perceived expectations. Much like how quantum particles exist in a state of superposition until observed, individuals may consider multiple potential decisions that solidify into a single choice when action is necessary. This view encourages us to appreciate the complexity of decision-making processes, urging us to consider the myriad possibilities preceding a single outcome. By embracing this uncertainty, individuals and organizations can become more adaptable, fostering environments where innovative solutions emerge from unexpected avenues.

Artificial intelligence plays a crucial role in navigating the unpredictability of human behavior, offering a powerful tool to analyze and forecast social outcomes. Through advanced algorithms, AI systems can identify patterns and correlations within vast datasets, uncovering insights that might otherwise remain hidden. In this sense, AI serves as a modern oracle, capable of providing probabilistic forecasts that account for the inherent unpredictability of human nature. For example, AI can analyze social media trends to predict shifts in public opinion or consumer behavior, offering valuable foresight for businesses and policymakers. By merging AI with the principles of quantum uncertainty, we can gain a more nuanced understanding of social dynamics, enabling more informed decision-making in an ever-evolving world.

The interaction between quantum uncertainty and social decision-making also encourages us to rethink traditional notions of control and predictability. In a world where uncertainty is a constant, rigid plans and deterministic models often fall short. Instead, adopting a flexible approach that embraces uncertainty can lead to more resilient strategies. By acknowledging the limits of predictability, organizations can cultivate a culture of experimentation and learning, where failure is seen as an opportunity for growth rather than a setback. This shift in mindset aligns with the quantum view that uncertainty is not a problem to be solved but a fundamental aspect of reality that can drive creativity and innovation.

Exploring the Heisenberg Uncertainty Principle in social contexts challenges our assumptions about human behavior. It prompts us to question the extent to which we can truly predict others' actions, encouraging a more inquisitive approach to social science. By recognizing the parallels between quantum mechanics and social dynamics, we gain valuable insights into decision-making, paving the way for a deeper understanding of the world. This exploration not only enriches our knowledge but also provides practical tools to navigate the complexities of modern society, fostering a greater appreciation for the intricate dance of human interaction.

Probability Waves and Their Influence on Group Decision-Making

In the realm of quantum mechanics, the concept of probability waves presents a fascinating analogy for understanding how groups make decisions. At the subatomic level, particles exist in multiple potential states, defined by a probability wave that resolves into a specific state only when observed. This notion can be likened to group dynamics, where decisions are fluid and undefined until a consensus is reached. Just as probability waves encompass various possible outcomes, group decisions reflect a multitude of potential results shaped by the diverse perspectives and motivations of those involved. By embracing this fluidity, groups can enhance collaborative problem-solving, exploring a broad spectrum of possibilities before arriving at a final decision.

Integrating advancements in artificial intelligence into this framework allows for a more refined analysis of group decision-making. AI algorithms can simulate the intricate interplay of individual preferences, biases, and social influences that steer collective choices, similar to how probability waves capture the range of quantum states. By modeling these interactions, AI can detect patterns and forecast possible outcomes, providing valuable insights into how groups might resolve disagreements or reach consensus. This predictive ability is particularly beneficial in areas like corporate strategy or public policy, where understanding the probabilistic nature of decision-making can lead to more strategic and informed choices.

Recent studies underscore the parallels between quantum uncertainty and social dynamics, suggesting that both fields' inherent unpredictability can inspire innovative decision-making approaches. For instance, just as quantum systems are sensitive to initial conditions, group decisions can be swayed by how information is presented or the order in which options are considered. Recognizing these subtle influences enables leaders to design decision-making processes that are more inclusive and representative of the group's diverse viewpoints. This perspective encourages a dynamic approach to leadership, shifting the focus from directing outcomes to creating an environment where the most promising possibilities can emerge.

Imagine a team tasked with developing a new product. Applying the concept of probability waves, the team could initially explore a wide array of ideas without rushing to a single solution. This open-ended exploration fosters the emergence of innovative concepts that might be overlooked if the group prematurely narrows its focus. AI tools can enhance this process by evaluating the potential of different ideas and suggesting combinations that maximize creativity and feasibility, guiding the team toward a decision that leverages the collective intelligence of its members.

Organizations can practically apply these insights by adopting decision-making frameworks that view uncertainty as a source of opportunity rather than a constraint. Encouraging diverse input and iterative feedback loops allows for the exploration of multiple futures before settling on one, much like quantum probability waves. By fostering an environment where the fluid nature of decisions is acknowledged and valued, teams can tackle complex challenges with greater agility and innovation. This approach not only improves the decision-making process but also nurtures a culture of continuous learning and adaptation, essential in today's rapidly evolving world.

The Role of Quantum Fluctuations in Predicting Social Outcomes

Quantum fluctuations, a fundamental aspect of particle physics, illuminate the unpredictability found in social dynamics. Just as particles behave randomly at the subatomic scale, social systems can undergo unexpected shifts that defy standard predictions. This randomness is evident in phenomena like abrupt swings in public opinion or the swift emergence and decline of social movements. By exploring these dynamics through the lens of quantum mechanics, we gain a richer understanding of the intricate dance between individual actions and collective outcomes. Recent research has shown that quantum models can adeptly capture the inherent randomness in social systems, providing fresh insights into the chaotic nature of human interactions.

In social decision-making, the probabilistic nature of quantum fluctuations challenges the deterministic frameworks of traditional models. By embracing the inherent uncertainty of these fluctuations, we can create more refined frameworks for forecasting social behavior. For example, the sudden surge of a trend or the unexpected triumph of a political campaign can be seen as akin to a quantum event, where minor initial variations lead to dramatically different results. Researchers are investigating how these principles can be applied to social networks, revealing that the interconnectedness of individuals can amplify these fluctuations, producing rapid, widespread changes in societal norms.

Artificial intelligence is crucial in leveraging quantum fluctuations for social forecasting. Advanced algorithms are being developed to incorporate these principles, allowing for more sophisticated analysis of social data. By considering the probabilistic and interconnected nature of social systems, AI can more accurately anticipate shifts in consumer behavior, political landscapes, and cultural trends. This fusion of quantum mechanics and AI not only enhances our predictive capabilities but also promotes a more adaptable approach to understanding social dynamics. As AI advances, its capacity to model these complex systems will inevitably lead to more accurate and insightful forecasts.

Applying quantum fluctuations to social theory invites us to reevaluate traditional notions of causality in human interactions. By recognizing the role of randomness and uncertainty, we open ourselves to a more comprehensive understanding of societal change. This perspective questions the linear models that have long dominated social sciences, suggesting that the future of social analysis lies in embracing complexity and unpredictability. By doing so, we can uncover patterns previously hidden by rigid frameworks, leading to more effective strategies for addressing societal challenges.

To fully grasp the potential of quantum fluctuations in predicting social outcomes, we must consider the broader implications of this approach. By adopting a particle physics perspective, individuals and organizations can become more adaptable and resilient in the face of uncertainty. Envisioning scenarios such as a sudden shift in public sentiment impacting policy decisions or market strategies encourages readers to apply these concepts in real-world contexts. By fostering a mindset that values flexibility and openness to change, we can better navigate the complexities of our social world, ultimately leading to more informed and effective decision-making processes.

Integrating AI with Quantum Uncertainty for Enhanced Social Forecasting

Blending artificial intelligence with the unpredictable nature of quantum mechanics presents an exciting chance to deepen our grasp of social forecasting. While AI excels at uncovering patterns in large datasets, it often struggles with the unpredictability of human actions. By incorporating quantum mechanics, particularly its principle of uncertainty, AI can create models that better reflect the probabilistic nature of social interactions. This synergy offers a more refined method for forecasting societal behaviors, recognizing that human decisions are not strictly deterministic but instead fluctuate within a range of possibilities.

Imagine applying quantum uncertainty to social networks, where individuals' choices and actions reverberate through their connections, forming intricate webs of influence. AI systems enhanced by quantum-inspired models can simulate these interactions with heightened accuracy, capturing the dynamic and often unpredictable nature of human relationships. For example, quantum algorithms could model the spread of trends within a community, accounting for the inherent unpredictability of personal choices and external factors. This approach not only offers a more comprehensive analysis of social dynamics but also unveils potential future trends that conventional models might miss.

The concept of quantum fluctuations—unpredictable variations occurring even in a vacuum—can be likened to unexpected events or decisions in social contexts that appear without clear cause. By merging AI with these quantum ideas, we can develop predictive models that accommodate the randomness intrinsic to social systems. This integration could transform fields like market analysis, political forecasting, and cultural trend prediction, where understanding the interaction of numerous unpredictable elements is crucial. Envision an AI that not only identifies current patterns but also anticipates sudden shifts, akin to a quantum state change when observed.

Moreover, the collaboration between AI and quantum uncertainty spurs innovation in real-time social forecasting. As AI evolves, its ability to adapt and learn from quantum-inspired insights will lead to more responsive, agile models that swiftly adjust to new data and contexts. This adaptability is vital in a world where social dynamics are in constant flux, driven by technological advances, cultural changes, and global events. By adopting this quantum viewpoint, researchers and analysts can craft tools that provide more flexible and accurate predictions, ultimately enhancing decision-making across various sectors.

It's important to consider the ethical and societal implications of integrating quantum principles with AI in social forecasting. How might these advancements impact privacy and individual agency? As AI becomes adept at predicting behaviors, balancing beneficial insights with potential overreach is crucial. Encouraging reflection on these issues can foster a deeper understanding of both the potential and limitations of this interdisciplinary approach, promoting thoughtful discussions on the future of social analysis in the age of AI and quantum thinking.

AI's Role in Analyzing Probabilistic Social Outcomes

Imagine a world where human behavior, once perceived as unpredictable and enigmatic, becomes as decipherable as a well-crafted puzzle. Here, the mysteries of social interactions transform into patterns, eagerly awaiting discovery. As we teeter on the edge of this new era, artificial intelligence emerges as a transformative tool, reshaping our understanding of social dynamics. This is not merely about calculations; AI interprets, drawing insights from the fluid principles of subatomic possibilities to illuminate the pathways of our collective decisions. By bridging the realms of physics and social behavior, AI offers a revolutionary perspective—one where the seemingly chaotic nature of societal choices aligns with the probabilistic dance of particles.

In this new landscape, machine learning serves as a guide through the intricate maze of human behavior, as AI systems develop models reflecting the inherent uncertainties of our social environment. The fusion of advanced physics concepts with AI ushers in an era of refined social forecasting, where future trends are mapped through real-time simulations, providing unprecedented insights into the rhythm of societal change. As we delve deeper, the potential for AI to model, predict, and even influence social outcomes becomes a tangible reality, inviting exploration into the profound implications of this technological evolution. In the pages ahead, we will journey through this captivating intersection, examining how these innovative tools are reshaping our understanding of human interaction in an increasingly complex world.

Leveraging Machine Learning to Predict Social Behavior Patterns

In the ever-evolving landscape of machine learning, predicting social behavior patterns presents a captivating blend of data, algorithms, and human psychology. Neural network-based models, in particular, have demonstrated impressive capabilities in extracting patterns from extensive and intricate datasets. These models analyze social media interactions, consumer behaviors, and digital footprints to anticipate trends and preferences. For example, sentiment analysis algorithms can assess public mood and foresee shifts in consumer sentiment, offering invaluable insights for businesses and policymakers. Leveraging the power of machine learning, it becomes feasible to predict social movements or economic changes, providing a strategic edge in various sectors.

One groundbreaking innovation in this area is reinforcement learning, a branch of machine learning where algorithms learn optimal behaviors through trial and error. This method is used to simulate social scenarios, allowing AI agents to interact within virtual environments and explore a multitude of social outcomes. Such simulations provide deep insights into potential human reactions under different conditions, enabling researchers to theorize about real-world social dynamics. By grasping these interactions, we can unravel the complexities of human decision-making and predict societal changes with greater precision.

Machine learning's strength lies not only in prediction but also in adaptability. Algorithms can continuously adjust their models as new data flows in, ensuring that predictions remain pertinent and timely. This adaptability mirrors the fluid nature of social environments, where variables are in constant flux. The ability of AI systems to update and optimize in real-time allows for more accurate social forecasting, effectively capturing the subtleties of human behavior. As a result, stakeholders can take proactive measures in response to potential societal shifts, rather than merely reacting after the fact.

Integrating quantum principles into machine learning marks a frontier with transformative potential. Quantum-inspired algorithms, drawing on concepts like superposition and entanglement, provide computational efficiencies and fresh insights into probabilistic reasoning. These algorithms can better manage the uncertainties inherent in social dynamics, offering a more nuanced understanding of fluctuating trends. By combining quantum mechanics with machine learning, researchers aim to develop models that not only predict but also elucidate the underlying causes of social phenomena, fostering a deeper comprehension of the intricate web of human interactions.

Looking ahead to the future of AI-driven social predictions raises intriguing questions. How can these systems be aligned ethically with societal values? What role will human intuition play as AI continues to advance? As we navigate these considerations, several actionable steps emerge: fostering interdisciplinary collaboration, investing in transparent AI systems, and encouraging public discourse on AI's societal implications. Such initiatives can guide us toward a future where machine learning is a powerful ally in unraveling the complexities of human society, transforming our understanding of social behavior into a precise science.

Utilizing AI Algorithms to Model Uncertain Social Dynamics

Artificial intelligence is transforming our perception of social dynamics by simulating the intricate network of human interactions with algorithms skilled at managing unpredictability. Traditional models often fail to capture the fluidity of social behaviors, where numerous variables shift due to internal and external factors. AI algorithms, however, excel in this area by using probabilistic models that reflect the inherent unpredictability of human societies. These algorithms can process extensive datasets, uncovering patterns and correlations that might escape conventional methods. For example, Bayesian networks and Markov models, which can adjust probabilities with new information, offer a solid framework for forecasting social phenomena like changes in consumer behavior or political sentiment.

Exploring further, the fusion of AI with quantum-inspired algorithms introduces a revolutionary method for modeling uncertain social dynamics. Quantum computing principles, such as superposition and entanglement, provide an edge over classical methods by allowing for the simultaneous consideration of multiple potential outcomes. This quantum approach enables AI to better handle the complexity of social systems, where linear cause-and-effect relationships are rare. Researchers are currently examining how quantum machine learning models can surpass traditional algorithms in predicting complex social interactions, incorporating a wide range of variables, including cultural shifts and economic trends. By adopting these advanced models, AI can offer unparalleled insights into the fluid nature of social systems, providing a clearer picture of possible future scenarios.

As AI advances, its ability to simulate and predict uncertain social dynamics will grow, propelled by groundbreaking research and technological progress. A promising area of development is hybrid models that combine the strengths of both classical and quantum computing. These models aim to boost AI's capacity to process and analyze data in real-time, delivering more accurate and timely predictions of social behaviors. Such innovations could transform fields from marketing to urban planning, where understanding human behavior is crucial. By embracing these technological breakthroughs, researchers and practitioners can unlock new possibilities for navigating the complexities of social systems, leading to more informed decision-making and strategic planning.

To fully harness AI's potential in modeling uncertain social dynamics, it is essential to integrate diverse perspectives and methodologies. Engaging with interdisciplinary approaches that draw from sociology, psychology, and computational science can enrich the development of AI models, ensuring they are comprehensive and adaptable. By incorporating insights from various fields, AI can better capture the nuances of human behavior, accounting for emotional intelligence and cultural context. This holistic approach not only enhances the accuracy of AI predictions but also fosters a deeper understanding of the social fabric underlying human interactions, paving the way for more effective and compassionate solutions to societal challenges.

In practical applications, AI's capacity to model uncertain social dynamics offers tangible benefits for organizations and communities alike. By leveraging AI-driven insights, businesses can optimize their strategies, tailoring products and services to meet changing consumer needs. Similarly, policymakers can use AI to identify and address emerging social issues, crafting policies that are both proactive and responsive. As AI continues to refine its capabilities, the potential for real-world impact grows, offering a powerful tool for shaping a more equitable and sustainable future. Embracing AI's role in understanding social dynamics not only enhances our ability to predict and adapt to change but also empowers individuals and organizations to thrive in an increasingly complex world.

Integrating Quantum Concepts in AI for Enhanced Social Forecasting

The fusion of quantum principles with artificial intelligence marks a revolutionary advance in social forecasting. Central to this innovation is the idea that the probabilistic essence of quantum mechanics can significantly boost AI's capacity to model the unpredictable nature of human behavior. In contrast to classical models that depend heavily on deterministic algorithms, quantum-inspired AI systems embrace uncertainty as an integral component. This perspective allows for a more sophisticated understanding of social phenomena, addressing the intricate web of human interactions and the numerous factors at play.

A particularly promising aspect of this integration involves the use of quantum superposition. While traditional computing processes data points linearly, quantum systems can evaluate multiple states simultaneously. This ability is mirrored in AI, enabling it to explore numerous potential social outcomes concurrently, thus providing a broader perspective on possible future scenarios. For example, utilizing quantum-inspired algorithms allows AI to simulate a range of societal responses to policy changes, offering a spectrum of potential reactions rather than a single predicted result. This equips policymakers with a richer array of data, aiding more informed decision-making.

Recent research underscores the transformative potential of quantum-enhanced machine learning in social prediction models. Quantum neural networks, for instance, have demonstrated an ability to discern patterns within complex social datasets that conventional methods might miss. By incorporating quantum principles into these networks, AI systems can detect subtle correlations and emerging trends vital in fields such as economics or public health. These advancements not only improve prediction accuracy but also offer insights into the dynamics driving societal transformations, opening new pathways for proactive interventions.

Additionally, integrating quantum concepts into AI supports the creation of adaptive systems capable of real-time simulation of social outcomes. These systems can adjust dynamically to new information, akin to how quantum particles respond to observation, continuously updating forecasts as societal conditions change. This adaptability is invaluable in today's fast-paced world, where static models can quickly become irrelevant. By leveraging the fluidity of quantum mechanics, AI remains relevant and responsive, offering timely insights that can guide actions and strategies across various social domains.

Exploring this intriguing intersection of quantum physics and AI invites a reassessment of traditional approaches to understanding social dynamics. By accepting the probabilistic and interconnected nature of quantum systems, we can construct AI models that more authentically reflect the complexities of human society. This novel perspective not only challenges conventional wisdom but also inspires deeper exploration into technology's potential to illuminate the intricacies of social interactions. As these quantum-inspired AI systems evolve, they promise to transform our perception and prediction of the ever-changing landscape of human behavior.

Developing AI Systems for Real-Time Social Outcome Simulations

Advancements in the realm of smart technology have ushered in systems that can perform live simulations of social scenarios, offering a unique perspective on the intricate dynamics of human interaction. This transformative capability is driven by advanced algorithms designed to process enormous datasets swiftly, mimicking the unpredictable nature of human decision-making similar to the fluctuations observed in the subatomic world. These intelligent systems leverage neural networks and deep learning strategies to capture the complex network of social relationships, predicting potential developments in various situations. Through scenario simulations, they provide valuable insights into possible outcomes, aiding individuals and organizations in navigating the often uncertain terrain of social engagement.

Beyond mere forecasting, real-time simulations allow for the exploration of hypothetical scenarios, akin to experiments in particle physics where numerous possibilities remain open until observed. This innovative approach enables stakeholders to test theories and comprehend the cascading effects of specific actions within a social framework. For example, urban developers might analyze the impact of policy shifts on community interactions, while companies could anticipate consumer responses to new initiatives. The ability to explore these scenarios in real time offers a strategic edge, enabling more informed and proactive decision-making.

The potential of AI to simulate social scenarios is further amplified by incorporating algorithms inspired by quantum principles, which introduce a fresh layer of complexity. These algorithms draw from concepts such as superposition and entanglement, allowing for modeling scenarios where multiple elements and results are interlinked. This method offers a nuanced understanding of social dynamics, capturing the intricacies of human behavior that conventional models might miss. By adopting this subatomic perspective, AI systems can produce simulations that reflect the inherent uncertainty and complexity of social interactions, offering a comprehensive view of potential outcomes.

As these sophisticated systems continue to develop, they hold the promise of transforming fields ranging from economics to sociology, providing adaptable tools for the ever-evolving landscape of human society. The ability to simulate social scenarios dynamically empowers researchers and practitioners to engage with complex social phenomena, offering insights that were previously out of reach. This emerging capability highlights the transformative power of AI when combined with quantum concepts, paving new paths for understanding and influencing the social sphere.

To fully capitalize on these advancements, practitioners must stay abreast of the latest developments and foster an interdisciplinary approach that bridges AI, quantum theory, and social science. Encouraging collaborations across these fields will help refine these systems, ensuring they remain attuned to the intricacies of human behavior. The ongoing challenge lies in balancing technical prowess with ethical considerations as we leverage these potent tools to shape the future of social analysis and intervention.

How Quantum Probability Explains Fluctuating Social Trends

To truly understand how the unpredictable nature of quantum mechanics influences societal behaviors, it's essential to recognize the underlying uncertainty present in both fields. Much like subatomic particles that remain in a potential state until observed, human preferences and cultural trends often linger in a state of possibility, waiting for the right moment to crystalize into something tangible. This interplay of possibilities mirrors the concept of quantum superposition, evident in the myriad choices individuals face daily, each decision shaping the ever-evolving landscape of social norms. The quantum framework offers a fascinating lens through which we can view the unpredictable shifts in public opinion and cultural trends.

The enigmatic connection between minds becomes apparent when widespread changes in public sentiment ripple through society, akin to quantum particles affecting each other across distances. In our digital age, the observer effect amplifies this phenomenon, as public attitudes are perpetually influenced by observation and media coverage. Whether it's the swift rise of a social movement or the fleeting appeal of cultural phenomena, quantum coherence provides a model for understanding how disparate groups synchronize. By examining these parallels, we uncover a complex web of interactions that reveal profound links between quantum theories and human behavior. This exploration offers insights into the underlying patterns driving social trends, paving the way for deeper discussions on each captivating aspect.

The Role of Superposition in Diverse Social Preferences

The principle of superposition, a cornerstone of quantum mechanics, provides an intriguing lens through which we can view the diverse array of social preferences in human interactions. In the realm of quantum physics, superposition allows particles to exist in multiple states simultaneously until measured. This concept has parallels in social dynamics, where individuals often hold an array of preferences and potential choices at once, shaped by various influences and contexts. These preferences are not fixed but fluctuate with new information, social interactions, and environmental changes. For example, consumer behavior can be seen as a superposition of potential choices, with preferences influenced by evolving trends, personal experiences, and social feedback. This dynamic nature challenges the traditional notion of linear decision-making, highlighting the need for a deeper understanding of how people navigate choices in their lives.

Recent advancements in artificial intelligence have given researchers the tools to delve into these complex layers of human preferences more effectively. Machine learning algorithms can simulate superposition by examining extensive datasets of human behavior, identifying patterns, and predicting shifts in preferences across different scenarios. These analyses reveal that individuals often juggle conflicting desires, much like particles in multiple states. Insights from such studies can be applied to fields ranging from marketing strategies that address diverse consumer needs to political campaigns that resonate with the multifaceted views of the electorate. By acknowledging the superposition of preferences, AI-driven models offer more personalized and dynamic solutions that reflect the true complexity of human decision-making.

This notion of superposition also illuminates the diversity and richness of human expression. Societal norms and values are not singular but are instead a superposition of historical influences, cultural exchanges, and individual interpretations. This perspective fosters a deeper appreciation for diversity, recognizing that cultures, like particles in superposition, embody multiple states and possibilities. Embracing this complexity can lead to more inclusive environments that celebrate differences while finding common ground. The interplay of diverse cultural elements can spark innovative ideas and creative solutions, akin to interference patterns observed in quantum experiments.

Digital platforms and social media play a critical role as observers in the quantum sense, collapsing potential states into specific preferences. These platforms curate content based on user interactions, often narrowing the range of available choices. This process underscores the balance between maintaining a diverse array of preferences and the tendency toward homogenization, prompting reflection on how digital ecosystems influence decision-making and encouraging users to remain aware of possibilities beyond algorithmic suggestions.

To apply insights from superposition practically, individuals and organizations can adopt strategies that acknowledge the fluidity of preferences. This might involve creating environments that encourage exploration and experimentation, allowing for the expression of multiple interests. Workplaces that promote interdisciplinary collaboration and diverse perspectives can tap into the creative potential of superposition, leading to innovative solutions and adaptive strategies. In personal growth, recognizing one's own superposition of preferences can foster greater self-awareness and resilience, helping individuals navigate the ever-changing landscape of choices with agility and openness. By embracing the quantum nature of social preferences, we can cultivate a richer, more dynamic understanding of human interactions and the myriad possibilities they present.

Entanglement and Its Influence on Collective Shifts in Opinion

In the realm of social dynamics, the concept of entanglement offers an intriguing analogy for understanding how collective opinions shift within societies. Much like the interconnectedness observed in quantum physics, this phenomenon reflects the complex web of human relationships where individuals and groups are often linked beyond direct interaction. When significant events occur—such as technological innovations, political changes, or cultural milestones—they send ripples through these networks, causing swift and extensive transformations in public sentiment. This interconnectedness reveals a hidden layer of social dynamics where individual beliefs and opinions are intertwined, resulting in synchronized shifts that can be as unpredictable as they are profound.

In social systems, entanglement can be likened to the way trends and movements gain traction. Individual choices are seldom made in isolation but are influenced by a network of social ties. For example, consider the rapid spread of social media campaigns or the sudden emergence of a cultural icon. These phenomena often arise because people are embedded in a social matrix where shared connections enable the swift dissemination of ideas and beliefs. As one person adopts a specific viewpoint or behavior, it can spark a chain reaction, encouraging others in their social circle to follow suit. This cascading effect exemplifies how entangled social settings can lead to significant shifts in collective opinion, often with a speed and magnitude that defy traditional prediction models.

Advancements in artificial intelligence provide promising tools to decode these complex networks. Machine learning algorithms, particularly those designed for network analysis, can map the intricate web of social connections and uncover patterns of influence and reciprocity. By simulating these interactions, AI can predict potential shifts in public opinion, offering valuable insights for policymakers, marketers, and social scientists. This computational approach enables a more nuanced understanding of how entangled relationships contribute to societal trends, providing a predictive edge in areas ranging from public health to political strategy.

Entanglement also challenges conventional notions of causality and linearity in social change. By acknowledging that collective opinion shifts are not merely the sum of individual decisions but rather the result of complex, intertwined interactions, we gain deeper insights into societal dynamics. This perspective encourages a reevaluation of strategies to influence public opinion, emphasizing the importance of targeting networks rather than individuals. Advocacy groups, for instance, might focus on influential nodes within social networks to initiate broader change, leveraging the entangled nature of social ties to amplify their impact.

Recognizing these interconnected dynamics prompts thought-provoking questions: How might this awareness alter our approach to conflict resolution or community building? What ethical considerations arise when leveraging entangled networks to influence public opinion? Exploring these questions invites readers to consider the potential of entanglement as a transformative lens for understanding and navigating modern society's complexities. This paradigm shift opens new avenues for engaging with social systems, highlighting the power of interconnectedness in shaping collective consciousness.

The Observer Effect in Media and Its Impact on Public Sentiment

In the mysterious world of quantum physics, the observer effect suggests that merely watching a phenomenon can change it. This concept finds a striking parallel in the influence of media on public sentiment. Much like an observer altering a quantum system, media shapes and sometimes sways public opinion through its presence and framing of information. This interaction highlights how media narratives can significantly impact societal trends, often amplifying certain issues and changing the course of public discourse.

Consider how economic forecasts are reported by news outlets. When media emphasizes potential downturns, it can lead to a shift towards public pessimism, triggering behaviors that may contribute to the very economic conditions feared. This scenario mirrors a quantum system collapsing into a defined state when observed. Here, media doesn't just report events; it actively shapes them, demonstrating the complex relationship between information and societal behavior.

Exploring the intersection of media studies and quantum mechanics opens new perspectives that challenge traditional views of communication. Researchers are investigating how the observer effect might be used to foster more informed and reflective public discourse. Recognizing media's dual role as observer and participant could lead to a media landscape that values transparency and accountability. Advanced technologies like AI-driven sentiment analysis provide tools to navigate this intricate environment, offering deeper insights into how media narratives influence public opinion.

Readers are encouraged to act as informed observers themselves. By critically engaging with media content and questioning how information is framed, individuals can reduce the observer effect on their own perceptions and choices. This active approach empowers people to navigate the media landscape with discernment, fostering a nuanced understanding of the relationship between media narratives and public sentiment. An actionable step could be to seek diverse information sources, gaining a comprehensive view of current events and minimizing media-induced bias.

As we consider the observer effect in media, we might wonder how future AI advancements could transform our understanding and interaction with information. Imagine AI predicting shifts in public sentiment before they occur, enabling proactive engagement with societal challenges. Such scenarios invite us to envision new ways of harnessing the observer effect constructively, enhancing our collective ability for informed decision-making and fostering a resilient, adaptable society.

Quantum Coherence as a Model for Synchronization in Cultural Fads

Quantum coherence, a phenomenon where particles align harmoniously, offers a compelling perspective to examine the rapid emergence and spread of cultural fads. Much like particles moving in unison, cultural trends often appear to arise spontaneously, capturing widespread attention and spreading quickly. This synchronization can resemble a resonant frequency, echoing through societal structures and leading diverse groups to adopt trends simultaneously. Viewing cultural fads through this quantum lens highlights the underlying patterns in how ideas and behaviors disseminate, moving beyond traditional models of social influence and diffusion.

The swift rise and decline of trends can be seen as a manifestation of quantum coherence, where individuals, similar to particles, align their choices and actions with those around them. This alignment is not random; it is influenced by various factors such as media, technology, and social networks. In today's interconnected environment, digital platforms act as catalysts, speeding up this process by enabling instant communication and feedback loops. This digital resonance fosters an atmosphere where cultural fads can thrive, often disregarding geographical or demographic boundaries. The coherence model thus explains how trends can quickly gain momentum and just as rapidly dissipate when collective attention shifts.

Recent studies in social physics are beginning to explore how quantum coherence might offer predictive insights into the lifecycle of cultural phenomena. By examining data from social media and other digital interactions, researchers can identify coherence patterns that signal the birth of a trend. These patterns, reflecting quantum states, are delicate and susceptible to external influences. A sudden shift in public sentiment, akin to a phase transition in quantum mechanics, can lead to the abrupt end of a fad. Understanding these dynamics allows for the anticipation of cultural shifts, providing valuable insights for marketers, sociologists, and policymakers aiming to harness or respond to these changes.

While the coherence model offers a fascinating framework for understanding cultural synchronization, it is crucial to acknowledge the nuances and complexities of human behavior. Not everyone will align with a trend, and dissenting voices can create interference patterns that alter its course. This diversity of responses underscores the importance of considering both the cohesive and discordant elements within a society. By incorporating these factors into their analyses, researchers can develop more comprehensive models that account for the full spectrum of human interaction and the unpredictable nature of cultural evolution.

By exploring the intersection of quantum coherence and social trends, a broader question emerges: how can individuals leverage this understanding to navigate the cultural landscape more effectively? By recognizing the forces that drive the spread of fads, individuals can make more informed decisions about their engagement with trends, balancing the desire for social alignment with personal authenticity. This awareness can also empower communities to foster cultural coherence that aligns with shared values and goals, contributing to a more harmonious and sustainable social environment. Through the lens of quantum coherence, the dance of cultural fads becomes not just a spectacle to observe but a dynamic process to engage with thoughtfully and strategically.

Reflecting on our journey through the realms of subatomic possibilities and their impact on collective human choices, this chapter sheds light on the delicate interplay between chaos and order that defines both particle physics and social interactions. By delving into the core significance of likelihood in quantum mechanics, we uncover striking similarities to the unpredictable nature of societal behaviors and movements. In this intricate landscape, artificial intelligence stands out as a formidable ally, adept at deciphering complex scenarios and offering profound insights into shifting social dynamics. This alliance between AI and quantum theories introduces a fresh perspective for interpreting the rise and fall of societal trends, highlighting the unexpected connections between distinct fields. As readers ponder these concepts, they are encouraged to rethink how randomness and chance influence their personal lives and the broader world. This introspection lays the groundwork for further investigation into how these subatomic ideas can enhance our grasp of collective human conduct, inviting a deeper contemplation on the evolution of society and the transformative potential of these insights for social analysis.

Chapter Nine

Quantum Field Theory And The Social Fabric

Have you ever stopped to ponder the elusive forces that bind our social world, much like threads weaving an intricate tapestry? These unseen influences guide us, connecting our lives in a vast network of interactions. Picture stepping into a lively city square, where every glance and word appears chaotic. Yet, beneath this surface, a pattern takes shape, crafted by the energy of human connections. This phenomenon echoes the mysteries of subatomic realms in physics, where hidden energies shape the behavior of particles, creating harmony from disorder.

As we delve into the wonders of these subatomic fields, we notice striking similarities to our social world. Just as these fields affect particles, unseen societal energies shape our interactions, subtly directing our choices and behaviors. These invisible currents influence how we form bonds, how trends arise, and even how communities evolve. By understanding these forces, we gain insight into the fabric of society, much like the cosmic dance that forms the universe.

The rise of AI offers a new lens to decode these societal currents. With its unmatched ability to process immense data, AI becomes a tool to observe and predict social trends with astonishing accuracy. In exploring this intersection of subatomic theories and AI, we uncover a powerful means to anticipate the flow of human behavior. By embracing this perspective, we deepen our understanding of social dynamics, empowering ourselves to navigate our interconnected world with greater awareness and purpose.

Quantum Fields: The Energy That Connects Everything

Take a moment to contemplate the hidden currents that compose the very fabric of our existence. These currents, akin to subatomic fields, are the mysterious forces that ripple through the cosmos, linking particles in a complex dance of energy and potential. Much like these subatomic fields, our societal realms are influenced by unseen energies that mold our interactions and relationships. Though intangible, these forces steer conversations, ignite trends, and forge communities. This intriguing parallel invites us to view human connectivity through a broader lens, revealing a universe where the social and subatomic are intertwined.

In exploring these connections, artificial intelligence becomes a formidable ally in charting the dynamics of our societal landscapes. By leveraging principles from particle physics, AI models can illuminate the intricacies of human networks, offering insights that might otherwise remain obscured. From forecasting the rise and fall of social movements to unraveling the development of collective behaviors, AI serves as a bridge between the subatomic and the social, uncovering patterns that reflect the fundamental forces of the universe. As we delve into these concepts, we discover the profound impact of subatomic fields on societal connections, unveiling a world where the unseen becomes apparent, and the abstract is made concrete.

The Fundamental Role of Quantum Fields in Social Connectivity

Quantum fields, though invisible, are crucial forces that intricately connect the universe, serving as a metaphor for the complexity of human relationships and networks. These fields, which dictate subatomic particle interactions, provide a lens through which we can understand the unseen forces shaping social interactions, akin to how energy fields influence particles over great distances. This analogy presents a fresh perspective on the subtle connections that weave the fabric of our social world.

Recent breakthroughs in quantum physics show that these fields are not static but are in constant motion, reflecting the fluid nature of human societies where relationships are always evolving. Just as particles respond to nearby energy fields, individuals and groups are shaped by their social environments. This dynamic interaction suggests that social connections, much like quantum fields, are more than just the sum of their parts, leading to emergent phenomena such as cultural shifts and societal trends.

Artificial intelligence is increasingly important in bridging our understanding of quantum fields and social connectivity. By using algorithms that mimic the probabilistic nature of quantum mechanics, AI helps uncover patterns in social networks that were once elusive. These models enable researchers to predict social phenomena with greater precision, offering insights into potential behaviors of individuals and groups under different conditions. Such simulations can reveal hidden links within social networks, deepening our understanding of how information, behaviors, and trends spread through societies.

An intriguing aspect of quantum field theory is its ability to explain emergent social phenomena—unforeseen outcomes that result from simpler interactions. In social contexts, these phenomena appear as trends or movements that suddenly gain traction. Quantum fields offer a framework for understanding how small, local interactions can escalate into large-scale changes. This perspective challenges traditional views of social dynamics, suggesting that the energy and influence within social fields can lead to nonlinear, unpredictable outcomes that defy conventional analysis.

To apply these insights practically, individuals and organizations can recognize the quantum nature of their social interactions. Understanding that relationships and networks are dynamic and interconnected allows for more effective adaptation to changing social environments. This awareness encourages a proactive approach to nurturing connections and anticipating shifts in societal trends. Fostering curiosity and open-mindedness can lead to innovative solutions and strategies, empowering people to navigate the complexities of modern social life with greater agility and insight.

Mapping Human Interactions Through Quantum Field Dynamics

In the intricate dance of human connections, the dynamics of quantum fields offer a transformative lens to view the unseen energies linking people and societies. Central to this concept is the idea that, akin to quantum fields, social links form a complex network of energy exchanges rather than isolated entities. This viewpoint suggests that each action and choice sends ripples through this social web, influencing outcomes in ways that aren't immediately apparent. Scholars are beginning to draw parallels between quantum fields and social interactions, proposing that the energy of human connections mirrors the fluctuating nature of subatomic fields, continuously shifting yet maintaining consistency over time.

The appeal of using quantum fields to map human interactions lies in their capacity to encompass the intricate and multidimensional aspects of social dynamics. Unlike conventional models that often simplify relationships into static or linear forms, quantum field dynamics embrace the complexity and unpredictability found in human behavior. By perceiving social interactions as dynamic domains, we gain deeper insights into phenomena like empathy, influence, and collective decision-making. Recent research has highlighted how quantum field principles can visualize the spread of ideas and emotions within networks, much like particles communicating through fields in physics.

AI technology has emerged as a vital tool in this exploration, excelling in modeling the complex patterns of social domains. Advanced algorithms analyze extensive datasets to uncover patterns akin to the intertwined states of quantum systems. For example, AI-driven models can forecast the spread of cultural trends or social movements by examining 'field potentials' within a community, including shared values, historical contexts, and underlying tensions. These models, informed by quantum field theory, provide a nuanced perspective for anticipating shifts in public opinion or societal norms. The integration of AI and quantum field concepts represents a cutting-edge intersection of technology and sociology, paving new paths for understanding and influencing human interactions.

Consider emergent social behaviors, which can be better understood through the quantum field theory perspective. Just as particles exhibit collective traits when interacting within a field, individuals can display emergent behaviors in society that aren't easily deduced from any single person's characteristics. These behaviors, whether seen as sudden cultural trends or spontaneous unity, can be viewed as outcomes of intricate field interactions. Applying quantum field dynamics to social analysis provides insights into how such phenomena emerge, offering a framework for predicting and potentially guiding these collective actions.

This examination of quantum field dynamics in social contexts encourages a reevaluation of the essence of connectivity in our world. By adopting a model that recognizes the fluidity and interconnectedness of human experiences, we can cultivate a deeper understanding of the invisible forces that shape our lives. As these insights intertwine with AI technologies, the potential for comprehending—and perhaps optimizing—our social interactions becomes limitless. This fusion of quantum physics and social science not only challenges traditional narratives but also equips us with the tools to navigate the complexities of modern society with greater clarity and foresight.

AI Models Bridging Quantum Fields and Social Networks

Quantum fields act as the unseen framework of the universe, intricately meshing the fabric of reality. Similarly, artificial intelligence has the capability to uncover the subtle threads of human interaction, highlighting the connections that unify us within social networks. Through advanced modeling, AI translates the abstract principles of quantum fields into the tangible world of social dynamics, offering profound insights into the complexities of human connectivity. By tracing the ebb and flow of interactions within these networks, AI can identify patterns that reflect the energy exchanges found in quantum fields, providing a rich tapestry of insights into the nature of our societal existence.

Visualize the social environment as a vast, interconnected web, where each person is a node and every interaction is a strand. AI excels in identifying the strength and direction of these strands, much like how quantum fields describe forces between particles. Recent advancements in machine learning algorithms have enabled AI to predict social movements and trends with remarkable accuracy by examining interactions within these networks. These algorithms sift through vast datasets of human behavior, detecting subtle shifts and emerging trends that might otherwise remain hidden. Such capabilities allow us to anticipate changes in societal norms, understand the rise of collective movements, and even forecast the trajectory of cultural phenomena.

Consider AI's application to social media platforms, where it continuously analyzes user interactions to map the intricate dynamics of the social field. By recognizing patterns in online behavior, AI can discern the formation of virtual communities and the spread of ideas, akin to tracking wave propagation in a quantum field. These insights can then be used to foster deeper connections, enhance communication strategies, and even curb the spread of misinformation. Such applications demonstrate AI's potential to bridge the gap between quantum field theory and social networks, providing a new perspective through which to understand our digital interactions.

Yet, this exploration is not without its challenges. The complexity of human behavior often defies simplistic categorization, requiring AI to adapt and evolve as it navigates the social field. Researchers are continually refining models to account for the nuanced and often unpredictable nature of social interactions. In doing so, they draw upon diverse disciplines, incorporating insights from psychology, sociology, and computer science to create robust frameworks capable of handling the intricacies of human connectivity. This interdisciplinary approach enriches our understanding of social fields, offering a more comprehensive view of the forces that shape our interactions.

As we delve deeper into this intersection of quantum fields and social networks, thought-provoking questions arise about the implications of these models. How might AI's ability to map social fields influence our perception of free will and agency? Could these insights be harnessed to foster more inclusive and equitable societies? By contemplating such questions, we are invited to reimagine the possibilities of human interaction and consider how these advanced models might be applied to create a more harmonious social fabric. This exploration challenges us to think critically about the role of technology in shaping our world and the ways we might harness its power to enhance our collective experience.

Emergent Social Phenomena Explained by Quantum Field Theory

In the intricate tapestry of our social world, quantum field theory offers a profound perspective to explore emergent societal phenomena. Similar to the unseen quantum fields that shape the physical universe, the social domain is an abstract but influential force guiding collective human behavior. This idea envisions society as a dynamic entity where individual actions, much like subatomic particles, both influence and are influenced by a larger interconnected framework. Scholars are increasingly exploring these parallels, revealing that subtle, often unnoticed interactions between individuals can generate complex societal patterns. While these may appear unpredictable on a micro level, they reveal coherent structures when viewed from a broader perspective.

One compelling example is the phenomenon of social synchronization, where spontaneous, widespread trends emerge across diverse populations. Quantum field theory explains how such patterns arise from interactions within the social realm, akin to particles aligning in a magnetic field. For instance, the rapid spread of cultural memes or viral trends illustrates a form of social coherence, where individuals' collective behavior aligns to form a unified movement. This alignment emerges organically, not through top-down control but through interactions within the social domain, highlighting the power of decentralized yet interconnected networks.

Artificial intelligence plays a crucial role in modeling these social areas, offering unprecedented insights into the mechanics of societal change. By utilizing vast datasets and sophisticated algorithms, AI can simulate and analyze social dynamics, uncovering patterns that might otherwise remain hidden. These models enable researchers to predict how minor disruptions, like quantum fluctuations, might ripple through a social network, leading to significant changes in collective behavior. These insights are invaluable for policymakers and social scientists seeking to comprehend and influence social dynamics in an increasingly complex world.

Incorporating quantum field theory into the study of social phenomena challenges traditional views of causality and influence. Conventional models often view social interactions through a linear lens, where cause and effect are straightforward and predictable. However, the quantum perspective suggests a more nuanced view, where interactions are probabilistic and outcomes depend on numerous factors. This has profound implications for how we approach social innovation and problem-solving. By embracing the inherent uncertainty and complexity of social fields, we can develop more resilient and adaptable strategies for fostering positive social change.

As we stand on the brink of a new era in social science, integrating quantum field theory provides a transformative framework for understanding the subtleties of human interactions. By viewing society through this lens, we gain a deeper appreciation of the interconnectedness of our social world and the tools to navigate and shape it more effectively. Encouraging exploration of these concepts invites a paradigm shift in how we perceive and engage with the social fabric, fostering a more nuanced and holistic approach to contemporary society's challenges and opportunities.

The Social Field: Invisible Forces That Shape Human Interaction

Recall a moment when you entered a room and sensed an indescribable energy—a silent force linking everyone in an unseen network. Whether it was the tension before a crucial decision in a meeting or the electric thrill at a live performance, these feelings are not mere illusions. They represent what might be called the societal field: a web of hidden influences that subtly shape human interactions. This concept parallels subatomic fields in physics, which are dynamic energies connecting particles across vast distances. Just as these fields form the foundation of the universe, societal fields provide the framework for our collective experiences, influencing how we communicate, collaborate, and live together.

Within this domain of societal fields, a complex dance of energies dictates group dynamics, similar to the forces that govern particle behavior. Social entanglement, akin to its quantum equivalent, becomes evident when individual choices weave together, impacting collective outcomes in surprising ways. As we explore this metaphorical landscape, we'll uncover how quantum resonance highlights the emotional connections that bind communities, while social superposition challenges our notions of identity and behavior. Each section reveals a layer of this intricate social tapestry, offering deep insights into the forces shaping our shared reality.

Unseen Energies in Group Dynamics

In the landscape of group dynamics, unseen influences often guide the flow of human interaction, much like the subtle forces that weave through the cosmos. These intangible elements, similar to the vibrational patterns in physics, can shape the delicate facets of social exchanges. When people gather, they form a network of influences that define the collective experience. This phenomenon is evident in various settings, from corporate meetings to community events, where mood, tone, and interaction outcomes are directed by these hidden social currents.

Recent research in social psychology highlights that these unseen forces are not mere byproducts of individual actions but emerge from the group as a whole. Studies have found that the emotional states of group members can align, leading to a shared emotional atmosphere. This mirrors the concept of quantum coherence in physics, where particles share a unified state. Such harmony in social groups can enhance decision-making, stimulate creativity, and boost performance as individuals resonate with the shared energy of the group.

The concept of social entanglement — akin to quantum entanglement — becomes clear when we observe how people's decision-making becomes intertwined in a group. This interconnectedness means that a change in one part of the system can impact the entire group, altering outcomes in unexpected ways. For example, a leader's enthusiasm can inspire a team, instilling purpose and direction. Conversely, negativity from one member can dampen the group's spirit. Recognizing these patterns allows us to use social entanglement as a tool for positive change and innovation.

Additionally, quantum resonance offers a compelling way to understand emotional connections within groups. Just as particles resonate at specific frequencies, individuals in a group can sync with certain emotional wavelengths. This resonance can foster belonging and understanding, creating a strong foundation for collaboration. By encouraging environments that promote positive resonance, leaders can build more cohesive and resilient teams, ready to tackle modern challenges.

Understanding these invisible social energies offers practical insights for enhancing group dynamics. By cultivating awareness of these forces, individuals can engage more consciously with the group's energy. Practical steps include promoting open communication, encouraging empathy, and creating spaces for reflection and feedback. By attuning to the subtle frequencies of social dynamics, individuals and organizations can unlock new potential, transforming how we connect and collaborate.

The Role of Social Entanglement in Collective Decision-Making

Human interactions often lead to a complex web of connections, much like the entanglement seen in quantum physics, where individuals become deeply intertwined through shared experiences, emotions, and goals. This interconnectedness influences collective decision-making, where one person's choice can send ripples through a network, affecting others in unexpected ways. Studies in social neuroscience and behavioral economics reveal how these unseen bonds synchronize group actions and thoughts. Picture a community united by a cause; the shared energy and vision create a synergy that drives collective action. These insights highlight the significant influence of social entanglement on group behavior, surpassing individual decision-making.

Social entanglement is not just a metaphor but a tangible phenomenon observable in various scenarios. During high-stakes negotiations or collaborative projects, group sentiment often sways decisions more than individual logic. This is evident in corporate settings where innovation thrives in teams with strong interpersonal connections, as members are more willing to take risks and support each other. This interconnectedness resembles quantum coherence, where aligned intentions boost group effectiveness. Emerging research in organizational behavior suggests that nurturing these connections can lead to teams that are more agile and resilient, capable of tackling complex challenges with ease.

Artificial intelligence provides tools to model and comprehend these social networks, uncovering the invisible forces at work. By analyzing communication patterns, emotional exchanges, and decision-making processes, AI can map the intricate social web within a group. This modeling predicts group dynamics and potential outcomes, offering valuable insights for leaders and policymakers. For instance, AI-driven simulations can help anticipate public reactions to policy changes or marketing strategies, enabling more strategic decisions. By leveraging these technologies, organizations can harness social entanglement to foster environments that support collective growth and innovation.

The intricate dance of social entanglement underscores the importance of empathy and emotional intelligence in decision-making. Those attuned to others' emotions navigate social landscapes more effectively, strengthening connections that enhance group coherence. This empathetic resonance is evident in leadership styles that prioritize emotional connectivity, leading to teams that are not only more productive but also more harmonious. As our world becomes increasingly interconnected, understanding and leveraging these relationships becomes a crucial skill. Experts in psychology and management emphasize cultivating empathy to strengthen social bonds and drive collective success.

Imagine a world where the principles of social entanglement are consciously applied to create cohesive and supportive communities. Encouraging individuals to recognize and nurture their connections can lead to more collaborative and inclusive societies. By fostering environments where social entanglements are valued and harnessed, we unlock the potential for profound social change. As we explore the intersections of quantum physics and social dynamics, the insights from social entanglement offer a blueprint for a future where human interactions are more aligned, purposeful, and transformative. This understanding empowers us to shape a society thriving on interconnectedness and shared purpose.

Quantum Resonance and Emotional Connectivity

Quantum resonance provides a vivid metaphor for grasping emotional connectivity within groups. In the realm of subatomic particles, resonance involves the amplification of waves when they align, creating a powerful synergy. Similarly, in communities and teams, emotional resonance emerges when individuals' feelings and attitudes harmonize, fostering a sense of unity and mutual understanding. This can be observed in various settings, from close-knit families and friendships to larger entities like workplaces and societal groups. As emotions resonate within these circles, they form an invisible yet potent bond, nurturing an environment where cooperation and empathy thrive.

Advancements in artificial intelligence offer a fascinating perspective on emotional resonance in social settings. Machine learning algorithms, adept at analyzing vast datasets, are beginning to identify interaction patterns that suggest a form of societal resonance. By examining communication styles, sentiment trends, and behavioral cues, AI can predict how emotional states might influence a group's dynamics. This insight holds significant value for fields like organizational psychology and network analysis, where understanding emotional connectivity can lead to better teamwork and more harmonious structures.

The concept of quantum resonance also highlights the delicate balance required to maintain harmony within a group. Just as quantum systems can be disrupted by external influences, collectives are vulnerable to discord when individual emotions become misaligned. Sustaining emotional connectivity hinges on fostering environments that prioritize open communication and emotional intelligence. Encouraging individuals to express their emotions and actively listen can help realign emotional states, restoring resonance and strengthening the group's collective resilience.

Exploring emotional resonance prompts intriguing questions about the nature of human connections. How do subtle shifts in mood or behavior ripple through a group, altering its core dynamics? Are there thresholds beyond which emotional resonance becomes irreversible, leading to either unity or fragmentation? These inquiries invite reflection on individual roles within social circles, considering how personal emotional states weave into the community's fabric and affect the resonance experienced with others.

To leverage the power of emotional resonance, individuals and leaders can adopt strategies to enhance connectivity within their groups. Promoting mindfulness practices, fostering inclusive dialogue, and utilizing AI-driven insights to understand group sentiment are ways to nurture resonance. By doing so, individuals can improve immediate relationships while contributing to a more connected and harmonious society. This approach aligns with the broader theme of viewing social interactions through a quantum lens, revealing the profound interconnectedness that underlies human relationships.

The Influence of Social Superposition on Identity and Behavior

The intricate relationship between identity and behavior is illuminated by the concept of social superposition, offering a sophisticated lens through which to understand how individuals juggle the varied roles they take on within society. Much like a subatomic particle existing in multiple states at once, people frequently manage overlapping identities shaped by cultural, familial, and personal influences. This fluid nature allows adaptability across different social settings, drawing on various aspects of their identities as needed. Recent research in social psychology highlights that this multiplicity is not merely a survival tactic but a core element of human interaction, fostering empathy and understanding across diverse communities.

Modern AI tools have begun to probe these complex identity layers with advanced algorithms that scrutinize social media data, communication styles, and behavioral trends. By analyzing how individuals present themselves on different platforms, AI uncovers patterns of social superposition, offering insights into how people reconcile conflicting identities. For example, a person might showcase a professional persona on LinkedIn while simultaneously expressing a more personal or creative side on Instagram. This duality underscores individuals' adaptability and resilience in managing multiple social identities—a concept AI continues to explore with increasing nuance.

The idea of social superposition also extends to group behaviors, where collective identities form and evolve. In any group, individuals bring unique perspectives, creating a mix of social energies that can lead to innovation and improved group dynamics. In corporate environments, teams that embrace diverse identities often outperform those that do not, leveraging a wider array of ideas and approaches. This phenomenon highlights the importance of fostering spaces where varied identities are acknowledged and appreciated, enabling organizations to thrive amid rapid change.

Emotional connections play a crucial role in social superposition, as individuals often resonate with others who share similar experiences or values. This resonance can create a kind of social entanglement, where one person's actions or decisions influence those around them, leading to collective shifts in behavior or opinion. Recognizing and leveraging these connections can empower leaders and policymakers to build more inclusive and cohesive communities by understanding the underlying forces that shape group identities and behaviors.

To tap into the potential of social superposition, individuals and organizations can adopt practices that promote self-reflection and open dialogue. Encouraging people to explore and articulate their multifaceted identities can enhance self-awareness and interpersonal understanding. Organizations might implement training programs that emphasize the value of diverse perspectives, fostering a culture of inclusivity and collaboration. By embracing the principles of social superposition, society can create environments where individuals feel empowered to express their full identities, leading to richer and more dynamic social interactions.

How AI Models Social Fields to Predict Movements and Trends

Human society can be likened to a complex mosaic, intricately assembled through myriad unseen connections and influences. These connections, reminiscent of the fundamental forces in the universe, exert a profound yet often unnoticed impact on our interactions. Within this elaborate network, artificial intelligence emerges as an innovative tool, capable of uncovering the hidden patterns that shape collective human behavior. By drawing parallels to the principles of subatomic physics, AI provides a novel lens through which we can observe and predict the currents of societal change. This approach not only deepens our comprehension of individual relationships but also equips us to foresee broader trends and shifts.

As we delve into the convergence of AI and subatomic principles in social modeling, a captivating spectrum of applications unfolds. Quantum entanglement offers a fresh perspective on analyzing networks by emphasizing the enduring connections between individuals, regardless of physical separation. Superposition sheds light on the fluidity of social identities, helping us to recognize the diverse roles individuals fulfill within their communities. The observer effect prompts us to refine predictive models by acknowledging how observation itself can impact outcomes. Meanwhile, ideas like quantum tunneling inspire fresh possibilities for predicting cultural transformations that challenge traditional expectations. Together, these themes weave a rich tapestry of insights, highlighting the dynamic interplay between subatomic physics and the social fabric, inviting readers to reconsider their understanding of the world.

Utilizing Quantum Entanglement for Social Network Analysis

Quantum entanglement illustrates how particles become interconnected despite being miles apart. This phenomenon serves as a powerful analogy for social networks, where human connections defy physical boundaries and significantly impact behaviors and decisions. By using the principles of entanglement, artificial intelligence can map these intricate relationships, offering new insights into how social ties affect influence and resilience in the face of change.

AI systems inspired by quantum concepts can uncover patterns in social networks that might otherwise remain hidden. They can pinpoint key influencers within these networks, much like entangled particles that have a significant impact on the entire system. This capability allows for the prediction of trends and shifts in public opinion. Advanced algorithms, informed by the complex nature of social interactions, can forecast how alterations in one part of the network may affect the whole, providing valuable insights for decision-makers across various fields. These tools can also reveal hidden connectors within the network—individuals or groups that play a critical role in the dissemination of information, akin to quantum particles that influence outcomes in unexpected ways.

The application of quantum principles extends beyond understanding existing connections to predicting future interactions. By analyzing entanglement dynamics, AI can foresee potential partnerships or disputes, offering a valuable resource for conflict resolution and community building. This predictive power arises from AI's ability to process vast datasets, uncovering subtle correlations that may elude human detection. Consequently, organizations can strategize more effectively, anticipating changes in social dynamics and adjusting their strategies accordingly. Such foresight enables more adaptable and responsive engagement with the social landscape.

Embracing the interconnectedness highlighted by quantum entanglement, AI-driven social network analysis can enhance understanding and empathy within communities. As individuals and organizations recognize the invisible bonds that connect them, they can make informed decisions that consider the broader impact on the network. This awareness can foster harmonious relationships and collaborative efforts, as parties acknowledge the mutual benefits of sustaining a balanced social ecosystem. The insights gained from this approach can also guide initiatives aimed at strengthening social cohesion and addressing issues like polarization and fragmentation.

As we explore the potential of applying quantum principles to social network analysis, it is crucial to consider ethical implications. Mapping and predicting social connections require a commitment to respecting privacy and autonomy. Balancing the use of these insights for positive outcomes with the protection of individual rights is essential. Promoting transparency and accountability in the application of AI models will ensure that the use of quantum concepts in social contexts remains beneficial, fostering a more connected and understanding society. Through careful application, the principles of quantum entanglement can illuminate the complexities of human networks, paving the way for more insightful and compassionate social interactions.

Applying Superposition to Understand Multidimensional Social Identities

Quantum superposition posits that particles can exist in multiple states at once until observed, offering a fascinating lens for exploring the intricacies of social identities. In our societies, individuals often embody a spectrum of roles and identities, each influencing behavior and interactions in complex ways. These can include professional roles, cultural ties, and personal beliefs, all woven into a dynamic tapestry. Viewing identities through the superposition concept allows for an appreciation of how people navigate these overlapping roles, which may seem contradictory or complementary depending on the context. For instance, someone might simultaneously see themselves as both a leader and a learner in different social spheres, highlighting the fluid and rich nature of identity.

Advanced AI models, inspired by quantum superposition, have transformed our capacity to analyze these multidimensional social identities. Machine learning can now process data from various sources—social media, professional networks, and consumer behavior—to build intricate profiles reflecting the complexity of an individual's identity. These profiles can foresee how individuals might act in certain situations, providing insights into decision-making processes that were once hard to understand. For instance, AI might anticipate a person's reaction to new social policies by considering their multifaceted identity, including political beliefs, cultural background, and economic status, thus offering a detailed understanding of societal trends.

The capabilities of these AI models go beyond mere prediction. They offer tools for deeper empathy and understanding in social interactions. By appreciating the superposition of identities, individuals and organizations can create environments that respect and integrate diverse perspectives. This can lead to more inclusive policies and practices where the multiplicity of identities is celebrated rather than constrained. In corporate settings, for instance, recognizing employees' multifaceted identities can improve team dynamics and foster innovation by drawing on a range of perspectives. Educational institutions can also benefit by designing curricula that reflect the diverse identities of students, promoting a holistic learning experience.

Intriguing questions arise when we consider the practical application of these insights. How might understanding superposition in social identities reshape conflict resolution strategies? Could it transform leadership styles to be more adaptive and responsive? By incorporating these questions into real-world contexts, individuals and organizations can navigate social complexities with greater agility. The recognition of superposition in social identities encourages a shift from binary thinking to a more nuanced approach that appreciates the full spectrum of human experience, fostering a culture of openness and adaptability.

To effectively harness these insights, practitioners can adopt specific strategies. Encouraging self-reflection and open dialogue within teams can uncover the underlying superpositions shaping group dynamics. In leadership roles, embracing flexibility and encouraging diverse viewpoints can lead to more innovative and effective decision-making. By applying these principles, individuals and communities can move toward a future where the richness of multidimensional identities is not only acknowledged but leveraged for collective growth and understanding.

Leveraging the Observer Effect to Refine Predictive Social Modeling

In the fascinating world of quantum physics, the observer effect demonstrates how observation itself can alter the phenomena being examined. This idea has an intriguing parallel in social behavior, where simply being watched can change how people act, often without them realizing it. In the field of predictive social modeling, artificial intelligence uses this concept to fine-tune algorithms, taking into account the variability introduced by observation. By incorporating this understanding, AI can more precisely predict social trends and patterns, acknowledging that the act of prediction might influence the outcomes it aims to foresee. This sophisticated approach highlights AI's capacity not only to analyze social trends but also to adapt to their evolution.

As AI develops, its models increasingly emulate the observer effect to boost the accuracy of social forecasts. Picture an AI system examining social media activity. It notes that users often adjust their behavior when they know they're being observed, similar to how people might behave differently during a recorded interview versus a private chat. By emulating this observer effect, AI can modify its predictions to account for behavior changes when individuals are conscious of their digital presence. This enhancement strengthens the reliability of insights drawn from data, ensuring predictions remain robust despite human variability.

This principle extends beyond individual behavior, impacting group dynamics and collective actions. In forecasting societal trends, AI must consider how public knowledge of predictive models might alter collective behavior. For example, if a community becomes aware that an AI predicts a rise in eco-friendly practices, this awareness itself might accelerate or impede that trend. This feedback loop requires a level of sophistication in AI systems, where predictions are constantly recalibrated as they become part of the social dialogue. The observer effect thus becomes a tool for refining AI's approach, embedding flexibility and responsiveness into its core operation.

Recent studies at the intersection of AI and social sciences reveal that integrating the observer effect can lead to more ethical and transparent predictions. By acknowledging the impact of observation, AI can be designed to minimize potential biases and unintended consequences, fostering trust and accountability. This approach encourages AI developers to consider diverse perspectives and cultural contexts, ensuring that predictive models are not only accurate but also equitable. By doing so, AI systems can provide insights that are both scientifically rigorous and socially responsible, bridging the gap between technological capability and human values.

To apply these insights practically, consider how the observer effect might be utilized in organizational settings. Companies could use AI models to forecast employee engagement trends while recognizing how transparency about these predictions might influence employee behavior. By openly discussing how predictions are made and their potential impact, organizations can cultivate a culture of awareness and adaptability. This proactive strategy allows for the creation of plans that anticipate changes and foster a dynamic environment, ultimately enriching both individual and collective experiences within the workplace.

Integrating Quantum Tunneling Concepts to Anticipate Cultural Shifts

Particles passing through barriers they seem unable to overcome, a phenomenon known as quantum tunneling, mirrors the abrupt cultural changes that can occur in societies. This challenges the notion that change always follows a predictable, incremental path, suggesting instead that cultural shifts can defy traditional expectations. By studying how ideas and trends can suddenly break through societal barriers, we can better understand the dynamic and unpredictable nature of cultural evolution. This insight helps us anticipate shifts in collective behavior, enabling us to navigate the complex landscape of human interaction more effectively.

Artificial intelligence, with its ability to process and analyze extensive data, is a crucial tool for modeling these cultural shifts akin to quantum tunneling. AI algorithms can uncover patterns within social data that might seem unconnected or minor individually. By pinpointing subtle indicators and correlations, AI can foresee when and how a cultural shift might happen, offering a snapshot of future societal landscapes. This predictive power is not just about forecasting

QUANTUM FIELD THEORY AND THE SOCIAL FABRIC   169

trends; it's about identifying the subtle signals that suggest a major change is imminent. In doing so, AI equips us to prepare for and potentially shape these shifts, aligning societal progress with desired goals.

Adopting quantum tunneling concepts in social analysis encourages a reevaluation of the barriers that ideas and cultural movements encounter. Traditional models often assume change is linear and constrained by obvious obstacles. However, a quantum perspective invites consideration of non-linear progress, where innovation and adaptation occur in leaps rather than steps. This reframing broadens our understanding of cultural adaptation, highlighting the need for flexibility and openness to rapid transformation. By embracing this mindset, individuals and organizations can become more resilient and better equipped to thrive in an ever-changing social environment.

Incorporating quantum tunneling principles into AI-driven social models not only enhances predictive accuracy but also fosters a deeper appreciation of cultural diversity. Recognizing that different societies and subcultures experience and express change uniquely allows us to avoid one-size-fits-all predictions. Instead, AI can be tailored to recognize the distinct traits that define various cultural contexts, enabling a more personalized and effective approach to anticipating cultural shifts. This sensitivity to diversity ensures that predictive models remain relevant and respectful of each community's unique dynamics.

Engaging with quantum tunneling in social dynamics invites us to consider how we might actively influence cultural change. Recognizing the potential for sudden, transformative shifts, we can adopt strategies that promote innovation and inclusivity. Thought-provoking questions arise, such as identifying and supporting emerging cultural movements or creating environments conducive to positive change. By applying these insights, individuals and organizations can proactively guide societal evolution, ensuring that cultural shifts align with broader goals of equity and progress. This approach enriches our understanding of the social fabric and empowers us to influence its ongoing transformation.

Reflecting on the intricate connections between quantum field theory and human society, this chapter has uncovered a compelling narrative of interconnectedness, where unseen energies subtly guide human interactions. By drawing parallels between subatomic realms and societal domains, we have explored how invisible influences shape behaviors and community trends. Through AI's lens, these societal patterns become clearer, offering unprecedented insights into human movements and trends. This exploration not only deepens our understanding of societal structures but also highlights AI's potential to harness these insights, potentially predicting and guiding the trajectory of collective human behavior. As we unravel the threads that bind us, we are invited to consider how these revelations might inspire a more harmonious and insightful social tapestry. With this fresh perspective, contemplate how the unseen forces in your interactions influence your understanding of community and change. As we move forward, the path is set for deeper reflections on how subatomic principles can further illuminate the complexities of human social evolution, challenging us to rethink our role within the societal fabric.

Chapter Ten
Quantum Energy And Human Motivation

Picture yourself in the heart of a lively city square, where the air buzzes with the pulse of daily life. Each person you pass, the lights that twinkle above, and the distant echoes of the city all form a complex symphony of vitality and movement. In this bustling environment, an unseen connection weaves through our intentions, akin to the energy levels that determine an atom's state. Just as electrons jump between paths, spurred by the energy they gain and relinquish, we too are driven by unseen forces shaping our actions and dreams. This chapter invites you to delve into this intriguing intersection where the microscopic world and our inner impulses converge in intriguing ways.

As we explore these ideas, consider how the rhythms of personal and shared energy shape our daily existence. How do our ambitions appear when viewed through the lens of subatomic energy? The similarities are striking, suggesting that much like particles influenced by their states, our drives are fluid, shifting with every internal and external influence. Seeing our aspirations through this lens offers a fresh outlook, proposing they are ever-changing, molded by countless interactions.

The rise of artificial intelligence introduces new avenues to comprehend these shifts. By leveraging AI, we can begin to gauge and even foresee the collective energy of humanity. This capability hints at the future of social analysis, where human behavior patterns can be unraveled with remarkable accuracy. As we embark on this chapter, envisioning ourselves as parts of a vast, interconnected tapestry becomes an exhilarating possibility. This journey not only deepens our understanding of what motivates us but also highlights the transformative power of viewing society through a quantum perspective.

Quantum Energy Levels: A Primer

Picture yourself amidst a boundless expanse, where every component constantly shifts, adapting to mysterious forces. This is the universe of subatomic energy states, a domain where particles engage in a seamless dance, transitioning fluidly between states—a ballet that embodies both chaos and harmony. Within this intricate choreography, we find reflections of our own lives, especially in the realms of personal drive and ambition. Just as particles transition between energy tiers, individuals navigate through varying levels of inspiration, influenced by a multitude of internal and external forces. Understanding these subatomic transitions offers a powerful metaphor for unraveling the complexities of human aspiration, providing a unique perspective on the ebb and flow of our desires.

As we delve deeper, the notion of quantum fluctuations becomes particularly captivating. These seemingly random shifts resemble the spontaneous changes in personal drive that can propel individuals and groups in new directions or cause unexpected stagnation. By drawing parallels between these subatomic phenomena and social interactions, we understand how minor changes in energetic states can lead to profound behavioral transformations. Here, artificial intelligence emerges as a remarkable tool, capable of measuring and predicting these shifts on a societal scale. As we explore these concepts, we'll discover how the principles of particle physics can shed light on the mysteries of human behavior, offering insights into the unseen forces that drive us and the potential for AI to anticipate and shape future societal patterns.

Exploring the Fundamentals of Quantum Energy States

Quantum energy levels, a fascinating aspect of modern physics, reveal that particles exist in specific energy states rather than a continuum. This concept underpins our understanding of atomic and molecular structures, showing that particles transition between these levels in defined steps. Such shifts require exact energy amounts, exemplifying energy conservation on a microscopic scale. This principle sharply contrasts with classical physics, which views energy changes as smooth and continuous, challenging our intuitive understanding of change.

Just as particles need specific energy to move between quantum states, our motivation can shift in response to new stimuli or realizations. This idea mirrors career growth, personal development, or societal changes, where a catalyst—such as inspiration, necessity, or opportunity—can elevate one's drive and ambition. Understanding this analogy helps us see motivational shifts not as gradual but as leaps triggered by distinct factors.

Recent studies on quantum fluctuations enhance our understanding of social dynamics, providing insight into the unpredictability of human interactions. These fluctuations, which represent temporary energy changes due to uncertainty, reflect the spontaneous and often unexpected nature of social behavior. This perspective helps explain how minor incidents can significantly impact social networks, leading to viral trends and shifts in public opinion, highlighting the complexity of human behavior.

AI has become a powerful tool in applying these quantum energy concepts to predict human behavior. Using algorithms that simulate principles of particle physics, AI can identify patterns within extensive datasets, revealing insights that were previously hidden. These discoveries can shape strategies to boost motivation within organizations, improve educational outcomes, or forecast economic trends. The integration of quantum physics and AI in social sciences holds transformative potential, enabling more informed decision-making and strategic planning.

To deepen our understanding of these intersections, consider how quantum energy concepts might inspire new models of leadership and organizational change. This exploration could lead to innovative approaches in team management, corporate culture, and personal development. By adopting this perspective, we not only expand our understanding of motivation but also unlock the full potential of human capabilities. These insights encourage us to tackle challenges with an open mind, driving meaningful improvements both individually and collectively.

How Quantum Transitions Mirror Human Motivational Shifts

Quantum transitions serve as a fascinating analogy for understanding the fluctuations in human drive and inspiration. Just as electrons move between energy states, releasing or absorbing force, human ambition similarly shifts in response to various stimuli. These transitions are not static; they evolve with changing circumstances, aspirations, and personal growth. By examining these changes through a quantum lens, we gain a richer insight into what propels individuals toward their goals or causes them to change priorities.

Consider the idea of quantum superposition, where particles exist in multiple states until observed. This complexity mirrors human drive, where multiple, sometimes conflicting desires coexist. The realization of a specific goal can focus one's efforts, just as observing a particle causes its wave function to collapse. This perspective highlights the fluid nature of ambition and the potential for sudden shifts in focus. Recognizing these parallels helps anticipate changes in drive and adapt strategies to stay aligned with personal and collective objectives.

Recent advances in AI offer powerful tools to map and predict these shifts in human drive. Machine learning algorithms, similar to those in subatomic computing, can sift through vast datasets to find patterns traditional methods might miss. These insights can forecast trends in ambition, providing valuable guidance in areas like organizational behavior, education, and personal growth. By applying principles from quantum transitions, AI can simulate outcomes of changes in ambition, enabling proactive strategies that align with emerging desires.

Practically, understanding how quantum transitions illuminate human drive equips individuals with a framework for personal growth. Viewing motivational states as dynamic encourages embracing change as part of one's journey. This mindset fosters resilience and adaptability, essential qualities for navigating an ever-evolving world. By setting flexible goals and reassessing priorities, individuals can use the energy of transitions to drive progress and innovation.

Engaging with the quantum perspective encourages deeper exploration of what drives us, prompting questions about the nature of desire and change. If ambition is inherently dynamic, how can we create environments that nurture positive shifts? What role do societal norms play in shaping these changes, and how can they be reimagined to promote collective well-being? These questions challenge conventional thinking and invite readers to apply quantum insights to their own lives, empowering them to effect meaningful change in their communities.

The Role of Quantum Fluctuations in Social Dynamics

In the complex realm of social interactions, the concept of quantum fluctuations serves as a powerful metaphor to grasp the unpredictable yet structured nature of human behavior. In quantum mechanics, fluctuations are brief changes in energy within a system, even when it seems stable. These shifts can be compared to the subtle fluctuations in social energy and drive that occur among individuals and groups—often unseen, yet deeply impactful. Just as minor quantum fluctuations can lead to observable physical changes, these social shifts can trigger significant behavioral transformations. By exploring these similarities, we gain a better understanding of the delicate balance between stability and change that defines human social environments.

Recent advancements in quantum social science suggest that just as quantum fluctuations can initiate phase transitions in physical systems, small disturbances in social settings can spark major shifts in group behavior. Imagine a workplace where a minor adjustment, like a new seating arrangement, sets off a chain reaction of interactions, reshaping the team's motivational dynamics. This parallels how quantum fluctuations influence energy states, propelling particles to new levels. By acknowledging these connections, organizations can leverage the understanding of quantum fluctuations to better anticipate and manage social dynamics, creating environments conducive to positive motivational shifts.

Utilizing artificial intelligence, researchers are now able to model these social fluctuations with increased accuracy, tapping into vast datasets to uncover patterns that might otherwise remain hidden. AI algorithms, akin to those in quantum physics used to predict particle behavior, can analyze social interactions and forecast when and how shifts in drive might occur. For instance, machine learning tools can monitor communication patterns within a team to predict potential conflicts or identify opportunities for enhancing collective drive. This blend of AI with quantum concepts presents a promising path for developing tools that not only predict but also guide social interactions toward more constructive outcomes.

While the analogy between quantum fluctuations and social dynamics is still being explored, it prompts intriguing questions about the nature of human drive. Are these fluctuations purely random, or do they follow a hidden, quantum-like order? This inquiry challenges traditional views, inviting us to consider the possibility that social behaviors may be less erratic than they seem. By applying quantum principles, we might discover underlying structures within social interactions, offering fresh insights into how drive can be influenced and guided. This perspective encourages leaders, educators, and individuals to rethink their strategies for engagement, focusing on the subtle, often overlooked factors that drive change.

Practically applying these insights involves adopting a mindset that embraces the unpredictable nature of social dynamics, much like accepting the inherent uncertainty in quantum systems. By developing an awareness of how minor disturbances can lead to significant outcomes, individuals and groups can learn to navigate social landscapes with greater agility and resilience. This approach not only deepens understanding but also empowers individuals to proactively shape their environments. Fostering an appreciation for the quantum nature of social interactions can lead to more adaptive, innovative strategies for motivating and inspiring others in a constantly changing world.

Applying Quantum Energy Concepts to Predict Human Behavior

Quantum energy ideas provide a compelling perspective to predict human behavior, drawing intriguing similarities between the fluctuating energy states of particles and the dynamic nature of human ambition. Observing these energy levels can offer insights into how individuals and collectives shift between states of drive, similar to electrons transitioning across energy tiers. This analogy underscores that just as particles need specific energy inputs to change states, human inspiration can be molded by various external and internal influences, such as social cues or personal ambitions. Identifying these influences allows us to better anticipate shifts in collective behavior, offering a model for predicting group responses to particular conditions or events.

Recent advancements in quantum mechanics reveal that quantum fluctuations can affect outcomes in unexpected ways. In the social realm, minor alterations in a group's environment or mood can lead to significant shifts in drive and behavior, mirroring these fluctuations. Consider the impact of a charismatic leader who, akin to a photon energizing an electron, can transform a team's motivational state, igniting a surge in collective enthusiasm and productivity. Applying quantum principles to societal contexts reveals how small, strategic actions can yield significant results, paving the way for more effective leadership and organizational change strategies.

Artificial intelligence plays a pivotal role in gauging and predicting these motivational shifts by analyzing extensive social data to detect patterns similar to quantum transitions. Advanced machine learning algorithms can capture subtle variations in language, sentiment, and behavior across digital platforms, offering real-time insights into group dynamics and potential shifts in drive. This capability enables organizations to adjust their strategies proactively, tailoring interventions to maintain or elevate group ambition, akin to adjusting the energy input to sustain a desired quantum state. By leveraging AI's predictive prowess, we can shift our understanding of human behavior from reactive to anticipatory, leading to more nuanced and effective engagement strategies.

Adopting a quantum view on motivation encourages consideration of individual variability and uncertainty in social dynamics. Just as quantum systems exhibit inherent unpredictability, human behavior is influenced by countless unpredictable factors. This perspective invites exploration of diverse motivational approaches, recognizing that what energizes one individual or group may not resonate with another. By embracing flexible, adaptive strategies that account for this variability, leaders and policymakers can create environments that effectively nurture drive and innovation, fostering resilience and adaptability in the face of change.

Exploring quantum energy concepts and their application to human behavior invites us to contemplate scenarios that challenge traditional thinking. What if, like particles in a quantum system, individuals possess untapped motivational states that can be activated through specific interventions? How might this potential be harnessed to create more dynamic and responsive social structures? These questions inspire critical thinking about the nature of drive and the possibilities for enhancing it through innovative approaches. By embracing this quantum-inspired perspective, we open ourselves to new possibilities for understanding and influencing the complex tapestry of human behavior, driving progress and transformation in our social world.

Human Energy and Motivation: A Quantum Perspective

Human ambition, akin to the dynamic nature of subatomic particles, is a constantly shifting force influenced by a complex array of factors. In today's world, where science and human experience increasingly intersect, examining personal drive through the lens of wave mechanics offers a captivating narrative. The similarities between quantized energy states and individual inspiration reveal a fascinating dance between the self and the collective, the conscious and the subconscious. Here lies a domain where traditional explanations fall short, giving way to a perspective that embraces complexity and unpredictability. This chapter invites readers on a journey to explore how our inner forces and aspirations resonate with the universal principles governing the particle world.

The entanglement of emotions within communities mirrors the delicate balance of subatomic states, where individual feelings can create waves of collective ambition. Just as particles can exist in multiple conditions simultaneously, humans can harbor conflicting aspirations and dreams, each representing potential paths. The observer effect, a compelling phenomenon in particle physics, finds its human counterpart in how our goals are shaped by awareness and perception. Moreover, the alignment of group energies reflects the coherence found in wave mechanics, where individual efforts converge towards a shared aim. Each subtopic peels back layers of understanding, offering a fresh perspective on how unseen forces guide our aspirations, energy, and collective momentum. Through this exploration, readers are encouraged to ponder the mysteries of human drive with renewed curiosity and insight.

Entanglement of Human Emotions and Collective Motivation

In the realm of quantum physics, entanglement refers to the intriguing connection between particles, where a change in the state of one particle instantly affects another, irrespective of the distance between them. This concept serves as a powerful analogy for understanding the complex network of human emotions and shared motivations. Much like quantum particles, human emotions can exhibit an invisible linkage to the feelings and aspirations of others within a social setting. For instance, when an individual experiences a surge of enthusiasm or creativity, it can influence the entire group, sparking a collective sense of purpose without direct interaction. This interconnectedness of emotions and motivations highlights the intricate and non-linear nature of human social dynamics.

Take, for example, a work environment where the mood and drive of one person can significantly affect the entire team's motivation. A leader's confidence and optimism can uplift the team's spirit, fostering an environment conducive to creativity and productivity. This mirrors quantum entanglement, where one person's positive energy resonates with others, encouraging a unified pursuit of common goals. This interconnectedness challenges the conventional view of motivation as an entirely personal endeavor, suggesting that our emotional states are closely tied to the social environments we inhabit.

Recent research in social neuroscience offers scientific support for this entwined view of human emotions. Studies show that mirror neurons in our brains play a key role in synchronizing emotions and motivations within a group. These neurons enable individuals to subconsciously replicate the emotional states of others, fostering a collective experience that can propel motivation. This biological basis of emotional entanglement underscores the need for nurturing positive emotional climates in settings where collaboration and shared motivation are essential.

With the advent of artificial intelligence, there are new opportunities to measure and analyze these intricate entanglements of human emotions and motivations. By employing sophisticated algorithms to process large volumes of data from social media, workplace communication, and other digital interactions, AI can identify trends and predict shifts in collective drive. This capability allows organizations to manage and enhance group dynamics proactively, optimizing motivation and performance. The interaction between AI and human behavior represents a cutting-edge approach to understanding and leveraging emotional entanglement for societal benefit.

As our world becomes increasingly connected, grasping the entanglement of human emotions and collective motivation is vital. By acknowledging the significant impact that individual emotional states can have on group dynamics, we can deliberately create environments that foster positive emotional connections. This understanding empowers both individuals and organizations to shape their social landscapes intentionally, maximizing potential for collective success. Thus, the quantum perspective not only deepens our insight into human interactions but also offers practical strategies for enhancing motivation and cohesion in our social endeavors.

Superposition States in Human Decision-Making and Potential Energy

Within the fascinating domain of quantum physics, the idea of superposition suggests that particles can be in several states at once until they are observed. This captivating principle provides a novel perspective for understanding human decision-making and the hidden possibilities within each person. Consider someone at a crossroads, such as choosing a career. Much like particles in superposition, this person juggles multiple options, each representing a different potential future. These possibilities coexist until a choice is made, narrowing the field to a single outcome. This analogy highlights the complex nature of human decisions and the immense potential that resides in moments of indecision.

Delving deeper into this metaphor, one can see how the superposition of choices creates a landscape brimming with potential akin to a reservoir ready to be unlocked. This potential isn't merely theoretical; it signifies the capacity for personal growth, creativity, and transformation inherent in undecided choices. In a world characterized by rapid change, the ability to tap into this potential is vital. By embracing the uncertainty of superposition, individuals can develop a mindset open to a variety of possibilities, leading to innovative and adaptable solutions. This approach resonates with modern psychological research, which emphasizes that flexible thinking and openness to new experiences are key to achieving personal and professional success.

The observer effect in quantum mechanics, where observation changes a particle's state, can be likened to the impact of external influences on human motivation. Just as observing a particle fixes its state, societal norms and cultural expectations can shape an individual's decisions, often limiting their range of options. By recognizing these influences, individuals can become more aware of external pressures and strive to maintain a broader range of possibilities. This awareness fosters a more self-directed approach to decision-making, empowering individuals to pursue paths that align with their inner motivations rather than conforming to external demands.

In today's fast-moving world, leveraging artificial intelligence is crucial for navigating the intricacies of human decision-making. Advanced algorithms can unravel patterns in decision-making processes, revealing how superposition states manifest in various scenarios. For example, AI can pinpoint factors that cause decision paralysis—a state where too many options lead to inaction. By understanding these dynamics, individuals and organizations can devise strategies to overcome decision fatigue and enhance the decision-making process. This practical application of AI not only improves personal choices but also shapes organizational strategies, fostering environments that encourage creativity and innovation.

Embracing the superposition of human decision-making equips people with the tools to handle uncertainty with assurance. By recognizing the potential inherent in untaken paths, individuals can build resilience and adaptability, crucial qualities in an ever-evolving world. This perspective shifts the perception of indecision from a barrier to an opportunity for growth. As individuals learn to harness this potential, they empower themselves and contribute to a more vibrant and innovative society. Engaging with these ideas prompts readers to reflect on their decision-making processes, challenging them to explore the uncharted potential within their own states of superposition.

Observer Effect and Its Influence on Personal Drive and Ambition

The observer effect, a captivating principle from quantum mechanics, implies that mere observation can change the state of a particle. When translated into human motivation, this concept suggests that awareness—whether self-awareness or external observation—can significantly influence one's drive and ambition. When individuals realize they are being observed or evaluated, their motivation often changes, sometimes enhancing performance or altering behavior. This is not just anecdotal; psychological studies support the idea that perceived scrutiny can lead people to adjust their actions. The implications for personal ambition are significant, indicating that simple reflection or awareness of being observed can spark a transformation in personal goals.

In the intricate dance of personal development, the observer effect acts as both a catalyst and a mirror. On one hand, being watched can push individuals to excel, much like athletes who perform better when an audience is watching. On the other hand, introspection allows individuals to step back and assess their own motivations, leading to greater self-awareness and a reevaluation of personal goals. This dual nature of observation—both external and internal—illuminates the paths through which ambition is channeled and amplified. By shifting between these two forms of observation, individuals gain insights into their motivational forces, enabling targeted and effective growth strategies.

The workplace offers a vivid example of the observer effect in action. Employees aware of being evaluated often show increased productivity and creativity, motivated by the knowledge of being watched. This dynamic can cultivate a culture of excellence, where ambition thrives under the scrutiny of peers and supervisors. Yet, the observer effect isn't limited to external observation. With self-monitoring technologies like fitness trackers and productivity apps, individuals can observe themselves. By tracking progress and achievements, people can use self-observation to fuel their drive, creating a feedback loop that propels them toward their goals with renewed energy.

The observer effect also invites reflection on feedback's role in sustaining motivation. Constructive feedback, similar to an observer's gaze, can highlight areas for improvement and inspire individuals to refine their ambitions.

This interplay between observation and feedback creates an environment where motivation is not static but an evolving force. Embracing feedback as a form of observation helps individuals build resilience and adaptability, essential for navigating modern life's complexities. In this context, the observer effect becomes a tool not just for enhancing performance but for fostering a mindset of continuous growth and learning.

Applying the observer effect to everyday life encourages a proactive approach to personal and professional development. By being conscious of both external perceptions and internal reflections, individuals can strategically harness drive to achieve their aspirations. This awareness offers a roadmap for setting realistic yet ambitious goals, informed by the understanding that observation—whether self-imposed or external—can be a powerful motivator. To integrate this concept into daily life, individuals might set regular intervals for self-reflection, seek feedback from trusted sources, and remain open to growth's potential. Through such mindful observation, the observer effect transforms from a concept into a dynamic force that shapes ambition and drives success.

Quantum Coherence and the Synchronization of Group Energies

Exploring the connection between quantum coherence and the collective dynamics of group energies opens up a captivating perspective on human collaboration that defies conventional boundaries. Quantum coherence highlights how particles achieve a state of synchronized harmony, akin to how individuals in a cohesive team align their motivations and actions toward a shared purpose. This synergy is evident in high-performing teams, where a unified vision and mutual understanding foster an environment where individual energies merge, resulting in extraordinary accomplishments. The synchronization in such teams stems not only from shared goals but also from a profound connection, reminiscent of the entangled states of quantum particles where a change in one influences the whole.

Research in social psychology and organizational behavior underscores the significance of coherence in boosting group effectiveness. Studies reveal that teams with strong coherence not only excel in performance but also exhibit increased resilience when facing challenges. This mirrors the quantum scenario where multiple particles function as a single unit, suggesting that when team members are emotionally and cognitively aligned, their collective potential grows exponentially. Beyond mere productivity, coherent groups often spark innovation, as the seamless exchange of ideas becomes more natural when energies are in sync. The challenge lies in cultivating this coherence, especially in diverse groups where differing perspectives and motivations need to be harmonized.

Artificial intelligence provides promising solutions for measuring and enhancing group coherence. By examining communication patterns, emotional tone, and decision-making processes, AI can pinpoint elements that either foster or hinder synchronization. This data-driven approach empowers organizations to devise strategies that align individual drives with collective aims, nurturing an environment where coherence akin to quantum alignment can thrive. AI tools can facilitate real-time feedback, allowing team members to dynamically adjust their interactions and strategies, similar to quantum systems adapting to maintain coherence. Harnessing AI in this manner represents a new frontier in organizational development, revolutionizing group operations and success.

Reflecting on quantum coherence in social contexts also highlights leadership's role. Leaders who grasp the essence of coherence can act as catalysts for group synchronization, employing strategies that echo quantum principles. This involves fostering environments where open communication and trust are crucial, enabling individuals to channel their unique energies toward a common vision. Such leaders understand that promoting coherence isn't about enforcing uniformity but encouraging diversity within a framework of shared understanding. By doing so, they emulate the quantum principle where diverse states enhance rather than undermine unity.

As we consider the alignment of group energies through the lens of quantum coherence, intriguing scenarios unfold. Imagine organizations consistently achieving coherence, much like a well-tuned orchestra, each member contributing harmoniously to the larger symphony. What innovations might emerge from such environments? How might societal challenges be tackled if collective energies were directed with precision and purpose? These questions invite us to rethink the potential of human collaboration, urging us to apply quantum insights to our social frameworks. By embracing coherence principles, individuals and organizations can navigate the complexities of modern life with renewed purpose and energy.

Using AI to Measure and Predict Shifts in Collective Motivation

Imagine a world where the shifts in human ambition and drive can be measured and predicted as accurately as the weather. This isn't just a futuristic dream; it's an emerging reality where artificial intelligence and quantum computing join forces to unravel the mysteries of social trends and group behaviors. At the core of this exploration is the idea that human drive, much like subatomic energy states, is dynamic and influenced by countless factors. By adopting this perspective, we unlock new ways to understand how these fluctuations shape our societies. This chapter aims to explore how AI technologies can reveal these hidden patterns, offering a fresh perspective on the intricate dance of human collective behavior.

In this new terrain, quantum algorithms evolve beyond abstract concepts to become crucial tools for analyzing social sentiment, capturing the subtle shifts in public mood with remarkable accuracy. As we venture into predicting trends in human drive, the fusion of AI with quantum models offers a groundbreaking method to transform chaotic datasets into clear insights. Machine learning enriches this process, dissecting complex group dynamics to uncover the forces driving societal change. Predictive analytics plays a key role, suggesting new social paradigms on the horizon. Through this blend of advanced technology and quantum insights, we reimagine the potential to anticipate and respond to the evolving motivations that guide human interactions on a grand scale.

Leveraging Quantum Algorithms for Social Sentiment Analysis

Quantum algorithms represent a cutting-edge leap in the realm of innovation, transforming how we analyze social sentiment. Borrowing principles from quantum physics, these algorithms offer remarkable efficiency in processing large, intricate datasets, thereby enriching our understanding of human emotions and interactions. For instance, quantum-inspired variational algorithms dynamically optimize data processing tasks, enabling researchers to detect subtle mood shifts that traditional methods might overlook. This capability allows for real-time public sentiment analysis, offering profound insights into collective emotions. Such insights are invaluable for policymakers and businesses aiming to align their strategies with evolving societal trends and emotional landscapes.

One significant advantage of using quantum algorithms in sentiment analysis is their ability to handle multiple states simultaneously, much like capturing various potential outcomes at once. This reflects the complexity of human emotions, which often exist as a spectrum of intertwined feelings. By utilizing quantum algorithms, researchers can model and anticipate these emotional complexities, providing a richer depiction of human sentiment. This mirrors the efforts of pioneering research labs developing quantum-based models to decode the intricate nature of emotional data, thus offering a comprehensive view of societal moods.

Additionally, the concept of entanglement in quantum mechanics finds a fascinating application in understanding interconnected social phenomena. When applied to sentiment analysis, this idea allows for examining how different societal groups influence each other's feelings, even across significant distances or demographic boundaries. Quantum algorithms can uncover hidden correlations within social networks, revealing how a change in sentiment in one region may echo across the globe. These insights deepen our understanding of global sentiment flows and are crucial for anticipating international reactions to political, economic, or cultural events.

By integrating these quantum approaches with advanced AI models, their potential is significantly amplified. This fusion creates tools that not only analyze but also predict shifts in collective behavior. Feeding quantum-processed sentiment data into machine learning algorithms enables researchers to identify emerging trends and forecast how societal motivations might change. This synergy between quantum computing and AI marks a significant advancement in predictive analytics, allowing for more precise and timely interventions in fields ranging from marketing to public health. As AI continues to evolve, its integration with quantum methods promises even more sophisticated analyses, hinting at future capabilities in social forecasting.

Imagine a scenario where organizations use quantum-enhanced sentiment analysis to anticipate changes in public opinion before they become apparent. What if governments could foresee societal unrest or economic changes through these predictive models, enabling preemptive actions to stabilize communities? These possibilities challenge existing paradigms, prompting us to consider the ethical and practical implications of such powerful tools. As we navigate this quantum frontier, the challenge lies in responsibly harnessing these insights to ensure they contribute positively to societal progress.

Integrating AI with Quantum Models to Forecast Motivation Trends

Artificial intelligence is becoming an essential asset in exploring human inspiration, especially when paired with quantum models. By merging AI's computational strength with the intricate principles of quantum mechanics, researchers are revealing previously hidden patterns in collective drive. Quantum algorithms, adept at managing the probabilistic nature of quantum states, offer a unique perspective to understand the dynamic and unpredictable aspects of human ambition. These algorithms enable AI systems to analyze vast datasets, detecting subtle changes in societal attitudes that might signal broader trends. As AI advances, its ability to predict these shifts becomes more refined, offering valuable insights into the fluctuating currents of public sentiment.

A particularly exciting development in this field is using AI to model and forecast trends in human drive utilizing quantum-inspired methods. These models transcend binary outcomes, embracing the inherent uncertainty and complexity of human behavior. By mimicking the superpositional states of quantum particles, AI can simulate multiple motivational states and estimate the likelihood of various outcomes. This approach allows for a deeper understanding of how motivations can evolve in response to external factors such as economic changes or cultural movements. For example, an AI-enhanced quantum model could anticipate a rise in environmental activism following a natural disaster, assessing the collective drive that might lead to significant societal transformation.

The fusion of AI and quantum models also provides a nuanced view of the interconnectedness between individual and collective ambition. Just as quantum entanglement describes the interdependence of particles, human motivations are often linked, influenced by social networks and cultural narratives. AI can map these complex webs of interaction, using quantum models to simulate how individual ambitions might resonate and amplify within a community, leading to widespread trends. This capability is particularly useful in understanding how grassroots movements gain traction and how societal norms change over time. It challenges traditional linear models of motivation, offering a more holistic view that considers the complex interplay of factors driving human behavior.

In practice, integrating AI with quantum models to predict trends in human inspiration involves training machine learning algorithms on datasets enriched with quantum variables. These datasets include not only traditional demographic information but also parameters that capture the probabilistic nature of human decision-making. Through iterative learning, AI systems refine their predictions, adapting to new data and evolving social contexts. This adaptability is crucial for accurate forecasting, as human ambition is inherently fluid and subject to rapid shifts. Researchers and organizations can use these insights to design interventions that align with predicted trends, fostering positive outcomes.

As we enter a new era of social analysis, the potential applications of AI and quantum models extend beyond prediction. They invite us to rethink the very nature of motivation, prompting questions about free will, social influence, and technology's role in shaping human desires. By challenging conventional paradigms and embracing the complexity of quantum mechanics, we open the door to innovative strategies for understanding and influencing motivation on a global scale. This integration encourages a deeper exploration of the factors that drive us, urging society to harness these insights for the collective good.

Utilizing Machine Learning to Decode Complex Group Dynamics

Exploring the potential of machine learning to unravel the complexities of group dynamics offers a transformative method for understanding collective human behavior. By employing sophisticated algorithms that mirror cognitive functions, machine learning analyzes extensive datasets to reveal the subtle patterns that govern social interactions. These algorithms, often inspired by neural networks, can discern trends from social media, forums, and other digital communication platforms, offering new perspectives on how groups form, evolve, and disband. This technology not only exposes the hidden layers of group behavior but also highlights the subtle shifts in collective sentiment that might otherwise remain unnoticed.

In recent years, researchers have leveraged machine learning to distinguish between types of group dynamics, such as cooperative versus competitive behavior, by examining the frequency and nature of interactions within digital communities. This distinction is vital for predicting potential outcomes of collective actions, like the success of a collaborative project or the fallout from a contentious public debate. By understanding these dynamics, machine learning can guide strategies for conflict resolution and enhance cooperative efforts, making it an invaluable resource for leaders and policymakers aiming to foster more harmonious and productive group environments.

One fascinating application of machine learning in this field is its ability to predict changes in group dynamics before they become apparent to human observers. For example, using sentiment analysis and natural language processing, machine learning models can identify early signs of discord or alignment within a group. These predictive capabilities are not merely theoretical; they have been applied in sectors like finance, where understanding group sentiment can lead to better market predictions, and in political campaigns, where gauging public opinion is crucial for developing effective strategies. The ability to anticipate these changes enables a proactive response to societal trends.

Furthermore, the intersection of quantum computing with machine learning opens new possibilities for analyzing group dynamics with unmatched precision and speed. Quantum algorithms can perform complex computations more efficiently than traditional methods, allowing for the analysis of intricate social phenomena involving numerous variables and dimensions. This collaboration between quantum computing and machine learning promises to unravel the complexities of human motivation on a scale previously considered impossible, paving the way for new approaches in social engineering and public policy development.

Consider the impact of these technologies in educational settings, where understanding group dynamics can enhance collaborative learning experiences. By analyzing student interactions, machine learning can determine optimal group compositions for diverse tasks, ensuring each member contributes effectively and learns from the experience. This tailored approach not only improves learning outcomes but also fosters a sense of community and belonging among students. The future of education, business, and society at large could be significantly shaped by these advancements, challenging us to continually refine our understanding of human interactions through the lens of cutting-edge technology.

Predictive Analytics and the Emergence of New Social Paradigms

Predictive analytics, when merged with quantum-inspired techniques, provides a revolutionary framework for forecasting changes in societal behaviors. By leveraging the intricate patterns identified through quantum algorithms, researchers can uncover complexities in vast datasets that traditional methods might miss. This innovative approach highlights hidden factors influencing group motivations, enabling the anticipation of social movements and transformations. The integration of quantum mechanics with predictive analytics offers a layered perspective, capturing the intricate nature of human societies. This analytical shift allows for a deeper understanding of how individual drives merge into larger societal trends, with significant implications for areas like economics and political science.

Imagine applying quantum algorithms to social media data, revealing subtle changes in public discourse that foreshadow major societal shifts. By viewing these changes through the concept of quantum superposition, researchers can interpret the multiple potential states of social sentiment, similar to particles existing in various forms. This view challenges the conventional binary data analysis approach, embracing the complexity of human emotion and thought. For example, an increase in discussions about climate change or social justice might herald a shift in public policy or consumer habits. Predictive analytics enables decision-makers to proactively adjust strategies, aligning with new societal trends rather than reacting to them after the fact.

Sophisticated machine learning models further boost these capabilities by unraveling the complex interconnections of group dynamics. These models, influenced by quantum entanglement principles, explore how individual actions resonate across networks, affecting collective outcomes. Understanding this interconnectedness is crucial for predicting the rise of new social norms or movements, as it acknowledges the ripple effects of personal agency within a collective framework. By simulating various scenarios, these models can project potential paths of societal evolution, offering invaluable insights for policymakers, businesses, and social organizations committed to fostering positive change.

Incorporating artificial intelligence into this framework not only enhances accuracy but also broadens the scope of predictive analytics. AI systems with quantum-inspired processing can analyze data streams in real time, providing immediate insights into changing motivational landscapes. This real-time analysis is essential for navigating the fast-paced changes of today's global society, allowing stakeholders to remain agile and responsive. Furthermore, AI can detect emerging trends and anomalies that may signal the emergence of new paradigms, offering a strategic advantage in foreseeing and shaping future societal developments.

As we enter a new era in social analysis, integrating quantum principles with predictive analytics urges us to reconsider the nature of societal evolution. What if the unpredictable, non-linear progression of social trends could be anticipated with greater precision? This possibility invites a reevaluation of strategic planning and decision-making processes across various sectors. By embracing this innovative approach, readers are encouraged to engage with the complexities of human interactions, recognizing the potential for AI and quantum models to not only predict but also inform and guide the emergence of new social paradigms.

In our investigation of the intricate relationship between subatomic dynamics and personal drive, we uncover a fresh lens through which to view the interplay of individuals and communities within the frameworks of society. By drawing parallels between the fluctuating states of quantum particles and the ebb and flow of human ambition, we achieve a more refined grasp of the forces shaping behavior. This analogy not only sheds light on how artificial intelligence might predict and quantify shifts in collective motivation but also offers a pioneering approach to social analysis. This fusion of ideas enriches our understanding of both individual initiative and group dynamics, highlighting the transformative potential of quantum-inspired perspectives in reshaping societal comprehension. As we venture onward, consider how the unpredictable shifts within our psyche reflect broader societal transformations and explore the opportunities that arise from this awareness. What novel insights might surface when these principles are applied to our daily interactions and the structures that bind us? This contemplation encourages readers to stay inquisitive about the unseen energies influencing our world and to actively engage with the transformative possibilities of a quantum viewpoint.

Chapter Eleven

Quantum Computing And The Future Of Social Analysis

Imagine a world where the intricate patterns of human society are as transparent and foreseeable as the pathways of subatomic particles. This vision is no longer the stuff of science fiction. Envision a bustling city where the movement of people mirrors the unpredictable yet patterned dance of particles in a quantum system. Within this chaos, patterns hide, waiting to be uncovered by the power of quantum technology. As we stand at the brink of a technological revolution, we must ask: how will these powerful machines, with their extraordinary capabilities, reshape our understanding of social interactions and dynamics?

At the core of this transformation lies the unparalleled ability of quantum systems to handle information in fundamentally novel ways. Unlike traditional computers bound by binary constraints, quantum systems use principles like superposition and entanglement to perform complex calculations with astonishing speed. This opens up the possibility of decoding the chaotic tapestry of human behavior, uncovering patterns that have long evaded social scientists. With this technology, the complexity of social systems becomes an opportunity rather than a barrier, revealing the hidden structures that shape our collective lives.

The combination of quantum systems and artificial intelligence further amplifies this potential, offering new dimensions of social analysis. As AI learns to navigate the quantum realm, it becomes adept at forecasting large-scale social behaviors, providing insights into everything from economic shifts to cultural trends. This chapter embarks on a journey to explore how these cutting-edge technologies might transform our understanding of the social world, weaving together the threads of quantum mechanics and AI to illuminate the profound links between the microcosm of particles and the macrocosm of human society. The implications are vast, inviting us to reimagine the fabric of social science in the light of this technological dawn.

The Basics of Quantum Computing: What Makes It Different

Quantum computing emerges as a transformative force poised to redefine the boundaries of computation and, consequently, reshape our understanding of society. Traditional computers operate on binary bits, locked in the state of either 0 or 1. In contrast, quantum computers leverage the mysterious capabilities of quantum bits, or qubits, which can simultaneously exist in multiple states. This phenomenon, known as superposition, empowers quantum systems to perform vast computational tasks in parallel, opening new realms of possibility. In a data-driven world, the ramifications of such advancements are significant, particularly in the field of social sciences. As quantum technology intertwines with artificial intelligence, the potential for sophisticated and predictive analyses of intricate societal patterns expands dramatically.

In this chapter, we embark on an exploration of quantum computing's core principles, inviting readers to uncover what makes this field distinct. A foundational understanding of qubits and their pivotal role in computation sets the stage. From here, we delve into the marvels of superposition, which facilitate unmatched parallel processing. The intriguing concept of entanglement offers insights into enhanced connectivity, potentially transforming our perception of relationships and networks. Finally, we examine the impact of quantum algorithms on data analysis, illustrating their capacity to revolutionize social science research. As we unpack these ideas, the chapter envisions a future where quantum computing and AI converge to provide unparalleled insights into the complexities of human interactions.

Understanding Quantum Bits and Their Role in Computation

Qubits, the fundamental units of quantum technology, mark a significant shift from the binary nature of traditional computing. Unlike classical bits that rest in states of 0 or 1, qubits can inhabit a superposition, existing as 0, 1, or both simultaneously. This unique property allows quantum systems to handle a vast amount of data concurrently, dramatically surpassing the abilities of even the most advanced classical supercomputers. The implications are profound, promising unprecedented efficiency and the potential to solve previously insurmountable challenges. As scientists improve the stability and coherence of qubits, the range of applications for this technology will expand, ushering in a new era of innovation.

Superposition not only accelerates computation but also transforms problem-solving strategies. By assessing numerous possibilities at once, quantum systems can explore vast solution spaces in parallel. This capability is invaluable in fields like cryptography, pharmaceuticals, and climate science, which involve complex optimization and modeling. For example, quantum systems might quickly crack encryption that is currently secure, necessitating a reevaluation of data protection strategies. This ability to process multiple outcomes concurrently offers a new perspective on computational challenges, redefining what can be achieved.

Entanglement, another fundamental principle, further enhances the power of quantum systems. When qubits are entangled, the state of one is instantaneously connected to another, regardless of distance. This phenomenon allows for unprecedented connectivity and synchronization between qubits, boosting the processing capabilities of quantum devices. In practice, entangled qubits can rapidly share and process information, potentially revolutionizing areas that require fast data transfer and analysis. As research progresses in maximizing and maintaining entanglement, its applications in quantum networks and secure communications are expanding, opening new avenues for exploration.

Quantum algorithms leverage these distinctive characteristics of qubits, offering methodologies that significantly outperform classical methods for certain tasks. Algorithms like Shor's for factorization and Grover's for searching highlight the computational superiority of quantum systems in specific scenarios. These algorithms not only emphasize the efficiency of quantum systems but also inspire new approaches to algorithm design. As developers experiment with quantum algorithms, the potential to discover new applications and enhance existing ones grows, providing fresh insights into the interplay between mathematics and technology.

Although this technology is still developing, its capacity to transform data analysis cannot be underestimated. By harnessing superposition, entanglement, and quantum algorithms, these systems promise unprecedented analytical capabilities. These advances are set to revolutionize social analysis, offering deeper insights into complex societal systems. As quantum systems integrate with artificial intelligence, the scope of understanding human behavior expands, providing novel insights into societal trends. The real challenge lies not in the potential of this technology but in effectively harnessing it to explore the complexities of human dynamics. This intersection of technology and social science foretells a future where computational power and human understanding converge in unprecedented ways.

Harnessing Superposition for Parallel Processing

Quantum computing sets itself apart through its unique ability to leverage superposition, a foundational concept where quantum bits, or qubits, can exist in multiple states at the same time. Unlike traditional bits that are limited to a binary state of 0 or 1, qubits can embody both simultaneously due to superposition. This feature revolutionizes computational power, enabling parallel processing on an extraordinary scale. In social analysis, this equates to processing vast datasets concurrently, revealing intricate patterns and correlations previously unattainable. Imagine analyzing entire social networks at once, gaining insights into how information and behaviors spread through societies with unmatched detail.

The practical benefits of superposition in quantum technology extend beyond speed. By evaluating multiple possibilities simultaneously, quantum systems can explore various scenarios and outcomes, making them invaluable for modeling intricate social systems. This opens new avenues for understanding phenomena like social mobility, economic fluctuations, and cultural shifts. Researchers, for example, could simulate the effects of policy changes on different demographic groups simultaneously, providing policymakers with a more comprehensive understanding of potential impacts. This marks a significant advancement from current methods that require sequential data analysis, often leading to oversimplified conclusions.

Recent studies underscore the transformational potential of superposition in tackling societal challenges. Research shows that superposition-enabled quantum algorithms can optimize resource allocation in social systems, such as healthcare and education, by modeling numerous variables at once. This not only enhances efficiency but also ensures a more equitable distribution of resources. For instance, by analyzing data on school performance, socioeconomic factors, and geographic distribution, quantum systems can propose optimized resource allocation strategies tailored to the specific needs of communities, ensuring maximum educational opportunities for all students.

The theoretical foundations of superposition also provide profound insights into human decision-making. Just as qubits can exist in a superposition of states until measured, individuals often consider multiple potential decisions before acting. This parallelism in thought processes can be mirrored in social analysis, where quantum systems facilitate the exploration of diverse outcomes based on varying initial conditions. Such an approach encourages a more nuanced understanding of decision-making, accounting for the complexity and variability inherent in human behavior. By embracing the probabilistic nature of quantum mechanics, researchers and analysts can move beyond deterministic models, acknowledging the rich tapestry of possibilities that shape social dynamics.

To fully leverage the potential of superposition in quantum technology for social analysis, a multidisciplinary approach is essential, bridging quantum physics, computer science, and social sciences. Encouraging collaboration across these fields will accelerate the development of innovative algorithms and methodologies that capitalize on the unique capabilities of quantum systems. This context raises thought-provoking questions: How might the integration of quantum technology reshape the landscape of social science research? What ethical considerations accompany the deployment of such powerful analytical tools? By pondering these questions, we pave the way for a future where quantum technology not only enhances our understanding of social phenomena but also enriches the human experience in profound and meaningful ways.

Exploring Entanglement for Enhanced Connectivity

Entanglement is a fundamental aspect of quantum mechanics that reveals a unique form of connectivity beyond what classical theories can explain. In the field of advanced computation, entanglement acts as a crucial mechanism for enhancing connectivity, allowing quantum bits, or qubits, to demonstrate a level of interdependence that traditional bits cannot achieve. This relationship enables qubits to exchange information instantly over any distance, offering significant implications for computational capabilities and efficiency. As scientists delve further into quantum entanglement, they discover new ways to leverage this phenomenon to tackle intricate problems that were once unsolvable by traditional computers.

One thrilling possibility of entanglement is its capacity to enable parallel processing on a scale that breaks conventional boundaries. By linking qubits through entanglement, quantum systems can execute multiple operations at once, vastly expanding their processing power. This parallelism is comparable to having numerous processors working in harmony, yet with a level of coordination unattainable by classical systems. As this technology advances, its potential to handle enormous data sets concurrently promises transformative impacts on areas like cryptography and pharmaceuticals, where analyzing complex molecular formations could lead to revolutionary breakthroughs.

The practical uses of entanglement extend beyond computation to include data analysis and secure connectivity. Envision a scenario where quantum networks use entanglement to establish secure communication channels immune to interception. Known as quantum cryptography, this concept exploits the distinct qualities of entanglement to ensure that any attempt to listen in would immediately alter the qubits' state, indicating a breach. Such developments not only enhance security but also build trust in digital communications, paving the way for more dependable interactions.

As exploration of entanglement progresses, researchers are also examining its potential in artificial intelligence and machine learning. Entangled qubits might enhance the connectivity within neural networks, improving pattern recognition and decision-making processes. By combining quantum technology with AI, we may see the rise of hybrid systems capable of addressing significant societal challenges, such as optimizing resource distribution or accurately predicting social trends. This merging of quantum advancements and AI suggests exciting possibilities for future innovations in social analysis.

Investigating entanglement in quantum systems prompts us to reconsider our understanding of connectivity and interaction in both digital and social contexts. Drawing comparisons between the interconnectedness of qubits and human networks offers new insights into relationships and collective behavior. This perspective encourages thoughtful consideration of how quantum principles could influence our approach to complex social issues, urging us to envision novel paradigms for connecting individuals and ideas in our increasingly interconnected world.

The Impact of Quantum Algorithms on Data Analysis

Quantum algorithms are revolutionizing data analysis by introducing groundbreaking methods for handling and interpreting intricate datasets. Unlike traditional algorithms that operate on binary digits, quantum algorithms utilize qubits, which can exist in multiple states simultaneously. This superposition enables the processing of vast data volumes concurrently, drastically enhancing computational efficiency. Consequently, quantum algorithms can address challenges that are beyond the scope of classical computing. A prime example is Shor's algorithm, which significantly speeds up the factorization of large numbers, demonstrating the transformative impact of quantum technology on cryptography and data analysis.

The unique properties of superposition and entanglement in quantum algorithms pave the way for new insights into data connectivity and pattern recognition. These capabilities allow algorithms to uncover hidden patterns and relationships within data that traditional methods might miss. Such advancements are particularly beneficial for social scientists aiming to decode complex social networks and dynamic interactions. Quantum algorithms can dissect intricate interdependencies and emergent behaviors within groups, offering a deeper understanding of collective phenomena. This potential has significant implications for fields like epidemiology, where tracking the spread of information or behaviors is crucial.

Furthermore, quantum algorithms offer a promising approach to redefining optimization problems common in the social sciences. Many societal issues, such as resource distribution and decision-making, can be viewed as optimization tasks. Quantum computing, with its ability to evaluate multiple solutions simultaneously, streamlines the search for optimal outcomes. Grover's Search algorithm exemplifies this by accelerating search processes beyond classical capabilities. These advancements could lead to breakthroughs in analyzing and predicting social trends, equipping policymakers with robust tools for informed decision-making.

As quantum algorithms advance, they are poised to reshape machine learning and artificial intelligence landscapes. The fusion of quantum computing with AI could elevate data analysis, enabling rapid and precise processing of complex, high-dimensional datasets. This synergy could result in sophisticated models capable of accurately predicting large-scale social behaviors. Leveraging quantum-enhanced AI, researchers could explore new dimensions in social forecasting, potentially revolutionizing how societies anticipate and respond to future challenges.

Incorporating quantum algorithms into data analysis encourages analysts and social scientists to reconsider traditional methods and explore innovative possibilities. The challenge lies in developing algorithms tailored to specific social science inquiries, fostering collaboration among physicists, computer scientists, and social researchers. As quantum technology evolves, staying informed about these advancements and considering their practical applications in social analysis is essential. Embracing the quantum frontier invites a reimagining of data analysis, encouraging a forward-thinking approach that questions conventional assumptions and embraces transformative change.

How Quantum Computing Will Revolutionize Social Science

Quantum computing is set to revolutionize the field of social science, offering a unique perspective to unravel the intricate web of human interactions. In our data-rich digital era, we navigate an immense sea of information that mirrors the intricacies of our societal exchanges. Yet, it is the quantum realm that holds the promise of unveiling the hidden patterns within this data, exposing a network of connections that traditional approaches barely scratch. Picture social networks as vibrant, dynamic ecosystems where every interaction sends ripples across the entire structure. Harnessing the power of quantum technology, which processes colossal volumes of data simultaneously, we can explore these ripples with unmatched clarity, revealing the complex tapestry of human connectivity.

As we delve deeper, the use of quantum simulations reveals its potential to enhance predictive models, marking a new frontier in comprehending human behavior. Quantum algorithms emulate the unpredictability inherent in human actions, akin to the chaotic dance of particles, opening doors to predicting societal trends with unprecedented sophistication. In this bold new era, quantum data analysis is poised to uncover societal patterns that have long eluded even the keenest observers. Quantum machine learning promises a transformative evolution in social science methodologies, shifting from rigid frameworks to more flexible, adaptive systems. Each aspect of this transformation invites us to reconsider the possibilities of social analysis in the quantum age, urging a reimagining of how we understand and interpret the complexities of human society.

Leveraging Quantum Algorithms to Decode Complex Social Networks

Quantum algorithms are revolutionizing our understanding of complex social networks by offering new ways to decode their intricacies. Unlike traditional methods, which process data linearly, these advanced algorithms leverage the principles of superposition and entanglement, enabling simultaneous analysis of multiple network pathways. This capability allows them to uncover hidden patterns and connections that might elude classical approaches. For example, in deciphering the dynamics of influence within a large organization, quantum algorithms can efficiently navigate vast relational data to identify key influencers, providing valuable insights into power structures and communication flows.

In social science, applying quantum algorithms to network analysis opens up possibilities for examining human interactions with greater precision. By treating social networks as quantum systems, researchers can simulate various scenarios, offering predictions of social processes with unprecedented accuracy. Such methods prove particularly effective where traditional models falter, like in forecasting the spread of information or behaviors within a community. This quantum framework facilitates a deeper understanding of how ideas propagate and social norms evolve.

Moreover, quantum algorithms offer a novel perspective on the resilience and adaptability of social networks. During crises or change, these networks face stress tests, and understanding their adaptability is crucial. Quantum approaches can assess the robustness of social ties and identify potential weak points that could lead to fragmentation. Such insights are invaluable for policymakers and social planners aiming to create resilient communities. By predicting how networks respond to external pressures, strategies can be developed to strengthen connections and promote social cohesion.

The integration of quantum algorithms with machine learning further enhances the analysis of social networks. Quantum machine learning accelerates the processing of complex datasets, making it possible to identify trends and anomalies at a pace previously unattainable. This capability is particularly relevant for analyzing online social platforms, where data generation occurs at high volumes and speeds, demanding more sophisticated analytical tools. As quantum computing technology advances, the synergy between these algorithms and machine learning promises to transform our understanding of digital social ecosystems.

To fully realize the benefits of quantum algorithms in social analysis, it is essential to address their ethical and practical implications. As these algorithms grow more adept at revealing deep-seated patterns in social networks, concerns about privacy and data security emerge. Establishing frameworks to ensure the responsible use of these technologies is vital, balancing individual rights with societal gains. Encouraging interdisciplinary collaboration and diverse perspectives, alongside prioritizing ethical considerations, can lead to more insightful and equitable outcomes in social science research.

Enhancing Predictive Models of Human Behavior with Quantum Simulations

Quantum simulations are reshaping the landscape of predictive models for human behavior, offering a novel perspective that surpasses the boundaries of traditional methods. Classical models often simplify human interactions into linear patterns, which can overlook the complex nature of social dynamics. In contrast, quantum simulations utilize principles of quantum mechanics, such as superposition and entanglement, to capture the intricate and multifaceted nature of human behavior. This approach allows for the examination of multiple possible outcomes at once, reflecting the probabilistic nature of human decisions. By integrating quantum principles, researchers can develop models that are not only more nuanced but also adaptable to the constantly changing social environment.

Beyond enhancing existing models, quantum simulations reveal previously hidden patterns in human behavior. These systems are adept at processing and analyzing vast datasets, identifying correlations that might be missed by classical approaches. For instance, when examining social media interactions or consumer habits, quantum simulations can quickly sift through massive amounts of data, uncovering detailed patterns that enhance predictive capabilities. Such insights could revolutionize areas like marketing, urban planning, and public health by providing a deeper understanding of subtle behavioral trends, leading to more effective strategies and interventions. This shift underscores the significance of adopting a quantum-focused approach in social sciences to unlock the full potential of data-driven decision-making.

Quantum simulations also bring advancements in modeling collective behaviors, such as crowd dynamics and market fluctuations, which are challenging to predict due to their inherent complexity. By leveraging the power of quantum simulations, researchers can model these systems with greater precision, considering their interconnected and emergent properties. This approach enhances prediction accuracy and provides valuable insights into the mechanisms driving group behaviors. In financial markets, for example, quantum simulations can help identify systemic risks and anticipate market changes, thus enabling more robust risk management and investment strategies. The capability to simulate such complex systems marks a significant leap in understanding collective behaviors.

Integrating quantum simulations into social science methodologies also prompts a reevaluation of ethical considerations and decision-making frameworks. As predictive models grow in sophistication, issues surrounding privacy, consent, and potential biases in simulations become increasingly important. The ability of quantum simulations to process large datasets necessitates careful consideration of data sourcing and usage. This calls for a balance between innovation and ethical responsibility, ensuring that the benefits of quantum simulations are realized without compromising individual rights or societal values. Engaging in this dialogue is crucial for creating a future where these technologies are used responsibly, fostering trust and transparency.

Imagine a future where quantum simulations empower individuals and organizations to make informed decisions, transforming how we tackle social challenges. By embracing this cutting-edge technology, researchers and practitioners can develop predictive models that are more accurate and reflective of our complex, interconnected world. The journey toward integrating quantum simulations into social science is not just about technological advancement; it is about reimagining the possibilities of human understanding and collaboration. As we stand at this new frontier, navigating the ethical, technical, and societal implications with foresight and creativity will be key to ensuring that this remarkable tool is used wisely and inclusively.

Unveiling Hidden Societal Patterns through Quantum Data Analysis

Quantum data analysis offers a revolutionary perspective for exploring the complexities of society, uncovering connections that traditional methods often miss.

Using the immense power of quantum technology, researchers can tackle the intricacies of social systems in ways classical computing simply cannot. Unlike conventional data analysis, which typically depends on linear predictions and simplified assumptions, quantum analysis embraces the complex, interconnected fabric of human societies. This approach enables a deeper comprehension of social phenomena, where variables interact in unexpected and synergistic ways, similar to the entanglement seen in quantum systems.

Quantum computers can handle vast amounts of data at once, identifying patterns and connections within social datasets that classical algorithms struggle to process. For example, social networks, with their complex webs of relationships and influences, can be more accurately mapped using quantum algorithms. These advanced techniques can sift through layered data to uncover the hidden structures that shape societal behaviors, such as how information spreads or cultural norms develop. Quantum data analysis sheds light on the subtle dynamics at play in these networks, offering insights into the proliferation of ideas and the emergence of collective behaviors.

Incorporating quantum simulations into social science research represents a paradigm shift, moving the focus from isolated variables to comprehensive systems. Through this lens, researchers can model entire social environments, exploring the potential outcomes of various societal interventions. This capability can greatly enhance predictive models, offering more precise forecasts of social trends and behaviors. For instance, by simulating the impact of policy changes on a community, policymakers can make better-informed decisions, anticipating unintended consequences and optimizing for desired societal outcomes.

The transformative potential of quantum data analysis lies in uncovering societal patterns that remain hidden under conventional scrutiny. By examining data through a quantum framework, researchers can detect anomalies and trends that signal shifts in societal attitudes or behaviors before they become evident in the mainstream. This proactive approach is invaluable for organizations aiming to stay ahead of cultural shifts, allowing them to adapt strategies in real-time based on emerging trends. As quantum data analysis becomes more refined and accessible, its applications in areas such as marketing, public health, and urban planning are likely to grow, driving innovation and more responsive strategies.

Engagement with quantum data analysis encourages reflection on the fundamental nature of societal dynamics and our methods of understanding them. If quantum insights suggest a reality where interconnectedness and uncertainty are fundamental, how might this influence our approach to social research? What new methodologies could emerge from embracing this perspective, and how might these reshape our understanding of human behavior? As we ponder these questions, the ongoing integration of quantum computing into social analysis promises to expand our cognitive horizons, challenging us to rethink the complexities of the social fabric and our place within it.

Transforming Social Science Methodologies with Quantum Machine Learning

Quantum machine learning is set to revolutionize social science methodologies by introducing unmatched analytical capabilities. This field leverages quantum computers' distinctive traits to handle vast datasets more efficiently than traditional approaches. This innovation holds the potential to transform how social scientists comprehend complex societal phenomena. By employing quantum algorithms, researchers can uncover intricate social patterns and relationships previously invisible. As quantum technology advances, it will enable more precise modeling of complex human behaviors and societal trends, opening new pathways for in-depth social analysis.

In their quest to understand societal complexities, quantum machine learning equips researchers to examine data with extraordinary detail. Unlike classical computing, which can be constrained by processing limitations, quantum systems can evaluate multiple possibilities simultaneously, speeding the discovery of subtle correlations and causations within social data. For example, quantum algorithms could revolutionize the study of social networks by revealing nuanced interactions and relationships that shape individual and collective behavior. This capability could lead to groundbreaking insights into the dynamics of social influence and cohesion.

The transformative potential of quantum machine learning extends to predictive modeling, a cornerstone of social science research. Quantum-enhanced models can accommodate a wider range of variables and interactions, producing more robust and reliable forecasts of social outcomes. In practical terms, consider public policy: quantum-enhanced simulations could offer policymakers deeper insights into the potential impacts of legislative changes on various demographic groups. By providing a more nuanced understanding of these complex interdependencies, quantum machine learning can inform more equitable and effective policy decisions.

Quantum machine learning also encourages a reevaluation of existing methodologies, prompting social scientists to adopt more innovative approaches to data analysis. Traditional methods often rely on linear models that may not capture the complexities of human behavior. In contrast, quantum approaches can incorporate non-linear dynamics, offering richer and more comprehensive models. This shift not only enhances the accuracy of existing research but also fosters the development of entirely new theoretical frameworks that better reflect the realities of human interactions.

As quantum machine learning progresses, it raises important questions about the future of social science research. How can we ensure that these powerful tools are used ethically and responsibly? What new ethical dilemmas might arise from these technological advancements? Social scientists must remain vigilant, balancing the potential benefits of quantum tools with considerations of privacy, fairness, and transparency. By fostering a dialogue around these issues, the integration of quantum machine learning into social science can be guided by principles that prioritize the well-being of individuals and societies.

AI, Quantum Computing, and Predicting Large-Scale Social Behaviors

In the dynamic fabric of human civilization, the convergence of quantum technology and artificial intelligence ushers in a groundbreaking era in deciphering large-scale social phenomena. Envision a world where the intricate web of human connections is decoded with extraordinary precision, allowing us to anticipate societal movements as clearly as meteorologists predict storms. This isn't just a distant dream; it's an emerging reality forged by the fusion of quantum advances and AI capabilities. As these fields progress, they promise to revolutionize our ability to interpret and forecast the rhythms of human collectives, akin to unraveling the mysterious dance of subatomic particles.

This transformative power is rooted in quantum algorithms that redefine computation itself, enabling the processing of immense datasets at speeds far beyond the reach of traditional methods. Quantum technology introduces a new paradigm in sociopolitical analysis, providing tools to unravel the complexities of collective behavior. The integration of AI with quantum systems opens new perspectives, allowing for the enhancement of predictive models that accurately capture the dynamics of societal trends and cultural transformations. As we delve into this journey, the chapter explores how these technological breakthroughs are set to reshape the field of social analysis, bringing us closer to a deeper understanding of our world.

Harnessing Quantum Algorithms for Sociopolitical Forecasting

Quantum algorithms hold the extraordinary promise of revolutionizing sociopolitical forecasting by offering new insights into complex societal systems. Unlike classical algorithms, these advanced methods can process massive datasets concurrently, revealing patterns and correlations that might otherwise be missed. By incorporating principles such as superposition, which allows data to exist in multiple states simultaneously, researchers can simulate and examine intricate scenarios that more accurately reflect the complexities of human behavior and decision-making. This approach fosters a deeper understanding of societal trends and potential developments, positioning this technology as a game-changer in social analysis.

Utilizing these algorithms for sociopolitical insights requires not only superior computational power but also innovative methods for interpreting the results. For example, employing quantum entanglement can help develop models that expose interactions between seemingly isolated societal components, echoing the interconnected nature of human communities. This ability is crucial for predicting electoral outcomes, shifts in public opinion, or the rise of new political movements. As this technology progresses, its integration with sociopolitical analysis could lead to more precise predictions and strategic planning, enabling policymakers to address challenges proactively.

The combination of these algorithms with artificial intelligence further amplifies forecasting capabilities. AI, known for learning from historical data and recognizing patterns, can be significantly enhanced by the processing power of advanced computation. This synergy allows for dynamic models that can adapt to real-time changes in societal data inputs. By applying machine learning techniques within this framework, researchers can create predictive models that not only foresee future trends but also uncover the underlying forces driving these changes, offering a comprehensive view of potential outcomes and their implications.

Applying these algorithms in this context also necessitates a reevaluation of traditional forecasting methods. Conventional models often depend on linear projections and historical trends, which may fall short in capturing the complexity of human societies. Quantum approaches, however, account for the uncertainty and unpredictability inherent in social and political landscapes. By considering multiple potential futures simultaneously, these algorithms provide a probabilistic range of outcomes, allowing decision-makers to prepare for various scenarios and adjust their strategies accordingly.

Looking ahead, the integration of these algorithms into sociopolitical forecasting represents an exciting new avenue for research and application. It challenges existing norms and encourages a rethinking of how we perceive and predict societal dynamics. By embracing this cutting-edge technology, researchers and policymakers can gain deeper insights into the mechanisms driving social change, paving the way for more informed and effective decision-making. This forward-looking approach not only enhances predictive capabilities but also enriches our understanding of the complex web of human interactions, fostering a more holistic and adaptive strategy to address contemporary challenges.

Unraveling Collective Human Behavior Through Quantum-AI Synergy

The fusion of quantum technology and artificial intelligence is reshaping our understanding of human behavior, surpassing traditional social science methods. Quantum systems, with their unmatched data processing capabilities, combined with AI's ability to detect patterns, offer a deeper dive into societal interactions. This collaboration deciphers the complex choreography of human actions, uncovering insights into group dynamics and collective mindsets. By operating with qubits instead of bits, quantum systems access previously unreachable data dimensions, capturing the nuances of human conduct with extraordinary accuracy.

Imagine a world where political trends and social movements are not just observed but predicted with stunning precision. When quantum algorithms merge with advanced AI models, they can simulate environments that unveil potential outcomes of societal and political scenarios. This predictive prowess goes beyond standard forecasting, providing a peek into the future of societal shifts. By examining variables through a quantum perspective, researchers can pinpoint patterns and correlations once deemed too intricate to comprehend, offering strategic advantages in policy and governance.

The combination of quantum technology and AI also unravels the complexities of collective human behavior by examining individual actions within broader social structures. Quantum concepts like entanglement resonate with social networks, where personal decisions and actions intertwine in intricate ways. AI can scrutinize these networks to detect emergent behaviors stemming from these interconnections, enhancing our grasp of phenomena like social contagion and group decision-making.

A quantum-AI framework promises advancements in understanding cultural dynamics and changes. As societies progress, cultural norms frequently shift in unpredictable directions. Quantum machine learning provides tools to model these cultural transitions, capturing the fluidity and interdependence of cultural components. By analyzing extensive cultural data, these models can forecast potential cultural trends and transformations, offering valuable insights for businesses, policymakers, and cultural organizations looking to stay ahead.

This exploration encourages reflection on the ethical and philosophical aspects of employing such powerful tools to decipher human behavior's complexities. As we extend the boundaries of knowledge, questions emerge about privacy, autonomy, and the limits of prediction. By contemplating these issues, we not only deepen our understanding of the potential of quantum-AI synergy but also engage in a broader conversation about technology's role in shaping humanity's future.

Enhancing Predictive Models of Social Movements with Quantum Computing

Quantum computing promises to reshape our understanding of social phenomena, offering insights that traditional computing struggles to match. Leveraging qubits, which can exist in multiple states at once, researchers can handle vast and complex datasets with incredible speed. This capability enables simulations of intricate social scenarios, capturing numerous variables and their interconnections to provide a nuanced view of societal dynamics. For example, simulating how a policy change might ripple through various societal layers can give policymakers a more comprehensive perspective on potential outcomes, leading to more informed and effective decisions.

One exciting application of quantum technology in social analysis is its ability to refine predictive models. Traditional models often simplify human behavior using linear assumptions and historical data. In contrast, quantum computing can incorporate nonlinear relationships and real-time data, capturing the fluidity and unpredictability of human societies. Envision quantum-enhanced models predicting shifts in public sentiment by analyzing extensive social media data, identifying subtle patterns and emerging trends overlooked by conventional methods. This could transform how governments, corporations, and NGOs anticipate and respond to social movements.

The synergy between quantum computing and artificial intelligence in the context of social movements offers groundbreaking possibilities. AI algorithms, powered by quantum processing, can analyze complex social phenomena with remarkable precision, identifying the catalysts that ignite movements, factors that sustain them, and conditions that lead to their decline. This understanding equips activists and organizers to strategize more effectively, ensuring targeted and impactful efforts. Additionally, predicting a movement's trajectory allows stakeholders to allocate resources wisely, manage risks, and engage with communities meaningfully.

As researchers delve deeper into the intersection of quantum technology and social science, new avenues for understanding cultural shifts emerge. Quantum-enhanced models can dissect the myriad influences driving cultural evolution, from technology and media to politics and economics. By exploring these interactions, we gain insights into the forces shaping societal values and norms. This knowledge is invaluable for educators, leaders, and innovators seeking to foster positive change in an ever-evolving world.

The potential of quantum computing to advance predictive models of social movements highlights the need for interdisciplinary collaboration. By bringing together physicists, computer scientists, and social scientists, we can develop sophisticated tools and methodologies that transcend traditional boundaries. This collaborative approach not only enhances our understanding of social phenomena but also equips society to tackle pressing global challenges. Embracing the transformative power of quantum computing paves the way for a future where social dynamics are not only understood but anticipated with unparalleled clarity.

Quantum Machine Learning and the Dynamics of Cultural Shifts

Quantum machine learning is revolutionizing our approach to comprehending and influencing cultural transformations. By leveraging advanced algorithms, researchers are charting new paths in social prediction, surpassing the limits of classical models. This technology's capacity to process enormous datasets efficiently unveils hidden patterns, allowing for simultaneous analysis of complex variables. Such capabilities offer a deeper understanding of cultural dynamics, shedding light on how societies evolve and change in response to diverse stimuli.

Consider cultural evolution, often likened to a complex, non-linear phenomenon where minor alterations can lead to significant transformations. Quantum machine learning thrives in this complexity, providing tools to simulate and predict the spread and evolution of cultural memes. Analyzing data from platforms like social media and news outlets, these algorithms detect emerging trends and shifts in public sentiment earlier than traditional methods. This foresight is invaluable for policymakers, businesses, and cultural institutions aiming to adapt to or influence these changes.

A captivating aspect of quantum machine learning is its ability to model cultural interdependencies. Cultures interact, influence, and sometimes clash; they are not isolated entities. Quantum models simulate these interactions, offering a comprehensive view of cultural dynamics. Understanding resonances or conflicts among cultural elements enables stakeholders to devise effective strategies for promoting cultural harmony or leveraging trends for innovation. This opens new possibilities in cultural diplomacy, marketing, and even conflict resolution, where grasping cultural undercurrents can lead to more successful outcomes.

Incorporating quantum machine learning into cultural analysis also redefines predictive models. Traditional approaches often struggle with the unpredictability of human behavior. In contrast, quantum models incorporate uncertainty as a feature, allowing for more flexible and adaptive forecasts. This paradigm shift encourages researchers to challenge assumptions about cultural predictability, adopting a dynamic, responsive approach. Consequently, they can provide robust predictions that account for the fluidity and complexity of cultural shifts.

Envisioning a future where quantum machine learning underpins cultural analysis prompts us to reconsider the essence of cultural change. If we could precisely anticipate shifts in cultural norms, how might this transform education, governance, and societal development? These questions inspire us to explore innovative possibilities and applications, urging us to transcend traditional frameworks. As quantum technologies evolve, they hold the promise not only of understanding cultural dynamics but also of actively shaping them, fostering a more interconnected and adaptable world.

The introduction of quantum technology into social analysis signals a groundbreaking era that promises to redefine social science with unmatched computational prowess and accuracy. By tapping into the distinctive properties of quantum systems, researchers gain the ability to decipher the intricate web of human actions and societal trends with greater clarity. This collaboration between quantum mechanics and artificial intelligence paves the way for more precise forecasts of large-scale social phenomena, offering deep insights into how individual decisions and collective behaviors intertwine. This technological leap not only expands our understanding but also arms us with better tools to foresee and tackle societal issues.

Reflecting on these groundbreaking innovations, we glimpse a future where the quantum perspective not only sheds light on the minutiae of subatomic realities but also enriches our grasp of the broader scope of human interactions. This chapter lays the groundwork for a wider conversation on how these insights will continue to influence our perception of social evolution and the dynamic interplay between individuals and communities. It encourages us to consider more deeply the role technology plays in shaping the future of human relationships. This exploration invites further reflection on the ethical and practical implications as we stand on the threshold of this new frontier.

Chapter Twelve

Entropy Disorder And Social Evolution

Amidst the gentle rustle of leaves, a hidden dance unfolds—a choreography of particles that echoes the unpredictable dynamics of human communities. Picture a world where every subtle change and every surge toward chaos is not mere chance but part of a grand symphony governed by fundamental principles. This chapter invites you to delve into the captivating idea of disorder and its impact on both the microscopic and macroscopic realms. Just as our universe gravitates toward complexity and unpredictability, so too does our social fabric, revealing an underlying order within apparent chaos.

Consider the vibrant scene of a bustling city, where the flow of human life seems erratic yet follows an unseen pattern. Here, the concept of societal disorder becomes tangible, illustrating how communities evolve, adapt, and sometimes unravel. In much the same way that physics describes a drift toward randomness, societies are shaped by invisible forces at work. The movements of people, the shifting of cultural norms, and the power of innovation all contribute to this dynamic landscape. By understanding these patterns, we gain insights into the delicate equilibrium between structure and disorder in our social world.

Within this interplay, artificial intelligence emerges as a transformative tool, capable of decoding the complexities of social change. Through advanced modeling and analysis, AI identifies the predictable threads woven into the fabric of societal evolution, offering glimpses into future transformations. As we journey through this chapter, the fusion of quantum insights with AI-driven analysis will not only illuminate the mechanics of societal evolution but also point us toward pathways for a more harmonious existence. This exploration invites us to reevaluate the forces shaping our world and to embrace the uncertainty that fuels growth and innovation.

Entropy in Quantum Physics: Moving Toward Disorder

At the crossroads of theory and practice lies a striking metaphor that bridges the microscopic and macroscopic realms: the concept of entropy. In the world of quantum physics, entropy quantifies the level of disorder, revealing a natural inclination for systems to drift towards chaos. This notion extends beyond abstract scientific principles; it reflects the intricate dance of stability and turmoil within human societies. Across the vast expanse of the universe, there is an inherent drift toward randomness, a principle that echoes through the tapestry of our social interactions. As individuals and communities navigate the shifting currents of societal norms, the forces of disorder manifest in evolving relationships, cultural shifts, and the relentless tide of social transformation.

This journey from the quantum scale to the societal sphere examines how disorder and the loss of information drive human evolution. Just as quantum systems confront the challenges posed by entropy, civilizations must grapple with the unpredictable elements that fuel change and adaptation. By exploring the similarities between quantum disorder and social dynamics, we gain a deeper understanding of the passage of time that influences both particles and people. As information dissipates into chaos, it leaves enduring marks on the trajectory of civilizations, prompting us to decipher and potentially forecast the path of our shared future. Through this lens, artificial intelligence emerges as a formidable tool, capable of modeling social complexity and offering glimpses of what lies ahead. As we delve into the subsequent topics, we will unravel these connections, tracing the threads that interlace quantum theory with the fabric of human existence.

The Role of Entropy in Quantum Systems and Social Dynamics

Entropy, a core notion in quantum theory, measures the level of chaos within a system and intriguingly mirrors the intricacies of human social interactions. In quantum contexts, entropy is not simply about disorder; it acts as a gauge for how systems shift towards equilibrium, harmonizing the unpredictable with the predictable. This refined understanding of entropy can deepen our insight into social frameworks, where interactions and relationships often seem chaotic but follow subtle, underlying patterns. By exploring entropy in quantum systems, we gain insights into the inherent tendency toward disorder that characterizes both atomic particles and social entities.

Consider how this concept appears in social dynamics: similar to particles in a quantum system spreading out and becoming less ordered, human groups naturally gravitate toward more diverse and complex configurations over time. This process can be observed in historical societal shifts, where cultural, political, and technological influences contribute to evolving social landscapes. The expansion of social networks, for instance, parallels quantum systems' movement toward higher states of entropy, as individuals form increasingly intricate and interconnected relationships. These dynamics highlight the dual nature of entropy, fostering disorder while also paving the way for new forms of order and innovation.

Recent strides in artificial intelligence have revolutionized our capacity to model social entropy, offering a novel perspective on the intricate balance between order and chaos in human interactions. By utilizing AI, researchers can simulate the evolution of social structures, providing a predictive lens through which we can anticipate future societal trends. These models draw parallels to quantum systems, where AI algorithms adeptly map the energy states and transitions of particles. This technological breakthrough allows us to view human societies as dynamic systems, continually adapting and reconfiguring in response to internal and external stimuli, offering a roadmap for navigating future complexities.

Exploring entropy in both quantum systems and social dynamics encourages a reevaluation of how we perceive stability and change. While entropy suggests a drift toward disorder, it also offers a canvas for creativity and growth. In social contexts, this implies understanding that periods of upheaval and transformation are not merely chaotic but are fertile grounds for innovation and development. This perspective invites a shift from viewing entropy solely as a force of destruction to recognizing its role in catalyzing evolution and resilience within societies. Embracing entropy as a natural and necessary component of social change empowers us to approach challenges with optimism and strategic foresight.

As we ponder the relationship between quantum systems and social dynamics, a compelling question emerges: How can we leverage the principles of entropy to foster positive change in our communities? By recognizing the patterns of disorder and order, we can identify strategic opportunities for intervention and innovation. For example, in organizational settings, leaders might use insights from entropy to design more adaptive and resilient structures that thrive amidst uncertainty. This approach transforms entropy from a seemingly abstract concept into a practical tool for social evolution, guiding us toward a future where complexity is not feared but embraced as a catalyst for progress.

Measuring Disorder: Entropy in Quantum Versus Classical Contexts

In the fascinating world of quantum theory, the concept of entropy plays a crucial role, representing the inevitable progression towards disorder within a system. While this idea is deeply rooted in subatomic interactions, it also offers significant insights when compared to classical interpretations of entropy. In classical physics, entropy is often seen as a measure of randomness or chaos in a macroscopic system, like a gas dispersing evenly in a room. Quantum entropy, however, operates on a more intricate level, capturing the probabilistic essence of quantum states and the uncertainty in predicting them. This distinction highlights a sophisticated understanding of how systems transition from order to disorder, challenging existing ideas about predictability and control.

Exploring further into the quantum realm, measuring disorder becomes more complex, especially when considering particle entanglement. Here, entropy goes beyond mere randomness, serving as a measure of the information lost when a quantum system moves from a pure state to a mixed state. This reflects how interconnectedness within quantum systems can lead to new properties that defy classical logic. For example, particle entanglement introduces a structured form of disorder, where the state of one particle is intrinsically connected to another, even over large distances. This idea shows how quantum systems can display a form of ordered disorder, where losing information unexpectedly results in new coherence across the system.

In modern research, investigating entropy in quantum contexts has led to innovative methods for measuring and understanding disorder. Techniques such as quantum tomography and quantum state reconstruction have become powerful tools for unraveling the complexities of quantum entropy. These methods allow researchers to map the probabilistic landscapes of quantum systems with remarkable precision, providing insights into the behavior of matter and energy at the most fundamental level. Such advancements not only strengthen theoretical models but also open the door for practical applications in quantum computing and information science, where managing and leveraging quantum entropy is key to developing resilient, error-resistant technologies.

The interplay between quantum and classical views on entropy raises intriguing questions about disorder in social systems. Viewing societies through a quantum lens reveals striking parallels: just as quantum systems exhibit entanglement and superposition, human interactions and societal structures often reflect interconnectedness and multiple potential outcomes. Measuring this social entropy poses a challenge, requiring innovative approaches similar to those used in quantum research. By employing artificial intelligence and machine learning algorithms, researchers can model social dynamics with a sophistication that mirrors quantum techniques, offering predictive insights into how societal order or disorder may evolve over time.

Considering potential applications, one could explore how understanding social entropy might inform strategies for managing change and complexity within organizations. Recognizing the inherent disorder and unpredictability in social systems can lead leaders to adopt more adaptive and resilient approaches to governance and decision-making. This perspective encourages a shift from rigid, deterministic models to those that embrace uncertainty and diversity, fostering environments where innovation and collaboration flourish. As we navigate the complexities of modern society, the lessons learned from quantum and classical interpretations of entropy hold transformative potential, inviting a reevaluation of how we perceive and respond to the ever-changing tapestry of human interactions.

Entropy and the Arrow of Time in Social Evolution

Entropy is a core idea in physics, acting as a link between the physical laws that govern nature and the intricate changes occurring in society. In quantum theory, entropy often reflects how disordered a system is, a notion that intriguingly parallels social structures. Much like a quantum system, societies naturally shift from order to disorder, influenced by cultural transformations, technological progress, and economic forces. This journey toward complexity and unpredictability aligns with the 'arrow of time' in physics, where systems evolve to increase entropy. Grasping social entropy involves understanding how communities move from organized, predictable states to more chaotic and dynamic conditions.

Recent research suggests that the concept of temporal asymmetry in physics can shed light on the evolution of societal frameworks. This one-way temporal progression indicates that once social systems change, they seldom return to their previous forms, akin to the ever-increasing entropy in physical systems. Take the digital revolution, for example, which has irreversibly altered how we communicate and share information, permanently reshaping societies. This irreversible nature of social evolution highlights the importance of recognizing patterns and forecasting societal changes. As entropy rises, societies often become more complex and interconnected, presenting new challenges and opportunities in governance, resource management, and cultural cohesion.

In social dynamics, entropy symbolizes the inherent unpredictability and diversity of human actions. Social entropy does not solely mean chaos; it also portrays the rich variety of human interactions and the multitude of possibilities within a community. By viewing social entropy through the lens of quantum physics, we gain insights into how minor, seemingly insignificant actions can ripple through a society, causing significant shifts over time. This perspective encourages us to consider how individual and collective choices contribute to the broader story of social evolution, similar to particles in a quantum field affecting the system's state.

The intersection of entropy and social evolution offers a fascinating prospect for AI to model and predict societal trends. Advanced algorithms, inspired by quantum principles, can simulate intricate social systems, providing insights into potential future scenarios. By analyzing extensive datasets, AI can detect patterns that might escape human attention, offering a more comprehensive view of how societies might evolve. This capability is crucial in anticipating societal responses to global challenges like climate change, economic instability, and technological disruption. As AI progresses, its role in understanding social entropy will likely grow, providing more accurate and actionable insights into the future of human interactions.

To leverage this understanding effectively, individuals and organizations could apply these insights to navigate the ever-evolving social landscape. By foreseeing changes in societal norms and values, businesses can adapt strategies to stay relevant, policymakers can develop more resilient frameworks, and communities can create environments that embrace change while maintaining core values. This proactive approach to social entropy enhances adaptability and empowers individuals to contribute positively to society's continuous evolution. Thus, the concept of entropy becomes not just an abstract scientific principle but a vital tool for comprehending and shaping the future of human civilization.

Quantum Entropy, Information Loss, and Societal Change

Quantum entropy, a key element in quantum theory, offers a refined perspective on the erosion of information in quantum systems and its broader impact on social evolution. Essentially, quantum entropy measures the unpredictability or disorder in a quantum framework, highlighting how information can become hidden as systems progress. This concept mirrors social contexts, where the dilution or misinterpretation of information can reshape societal norms and structures. As communities become more intricate, the scattering or degradation of information—akin to entropy—can affect social stability and adaptability. This connection between quantum entropy and social change suggests that just as particles transition to higher states of disorder, societies may also shift toward greater complexity and unpredictability, presenting both challenges and opportunities for innovation.

The idea of information decay in quantum systems serves as a metaphor for the fading collective memory within societies. Similar to quantum systems losing information through processes like decoherence, societies might see a weakening of cultural knowledge, traditions, and historical narratives over time. These losses can result from generational transitions, technological progress, or intentional attempts to revise history. This information decay can destabilize or propel societal advancement, depending on how communities manage and adapt to these shifts. By exploring this metaphor, we gain insights into balancing the preservation of cultural information with adopting new narratives that encourage social evolution. The challenge is leveraging this entropy to foster vibrant yet cohesive social frameworks capable of thriving amidst change.

Artificial intelligence stands out as a potent resource for modeling and predicting social disorder, offering unmatched capabilities for analyzing intricate social patterns. Through AI algorithms, researchers can simulate how information circulates and dissipates within communities, pinpointing potential vulnerabilities or points of transformation. These models can uncover hidden links between diverse social phenomena, demonstrating how minor changes in information dynamics can lead to significant societal shifts. For instance, AI can forecast the spread and influence of misinformation campaigns, helping policymakers craft strategies to counteract their effects. By understanding social entropy through AI, societies can better anticipate and navigate modern social evolution's intricacies, fostering resilience amid uncertainty.

The relationship between quantum entropy and societal transformation encourages a reevaluation of how information is perceived and interpreted within social frameworks. As quantum systems show, entropy is not solely a force of chaos but also a catalyst for complexity and innovation. Similarly, social disorder can be seen as a driving force for transformation, pushing societies to adapt, innovate, and evolve. By adopting this viewpoint, individuals and institutions can nurture a mindset that values flexibility, creativity, and inclusivity. Encouraging diverse perspectives and fostering open dialogues can mitigate the negative aspects of information decay, transforming potential chaos into a catalyst for positive change. This shift in mindset can empower communities to harness entropy's transformative power, leading to more resilient and adaptive social systems.

Reflecting on the parallels between quantum entropy and societal transformation, consider practical applications of this understanding. How can organizations and communities effectively manage information disorder to foster innovation and resilience? One strategy is to establish systems that prioritize preserving and accessing information, ensuring valuable cultural and historical knowledge is maintained and shared across generations. Additionally, creating environments that encourage experimentation and adaptability can help societies leverage disorder as a source of growth and creativity. By embracing the inherent uncertainty in social evolution, individuals and communities can transform potential chaos into opportunities for progress, crafting a future that is both dynamic and sustainable.

Social Entropy: How Societies Evolve and Change Over Time

Imagine a society, much like the vast universe, constantly finding its way between order and chaos. This intricate dance between structure and disorder is not just a key aspect of social evolution but also a mirror of the forces at work in human interactions. As communities expand and transform, they undergo changes reminiscent of quantum systems—dynamic, unpredictable, yet following certain patterns. Here, the idea of social entropy provides a fascinating perspective to observe the rise and fall of civilizations. It reflects a natural journey toward complexity, where established norms may dissolve, leading to new, sometimes surprising social formations. This process, akin to tendencies seen in quantum theory, shows how chaos can drive transformation, resulting in emerging societal paradigms.

At the core of this change is the exchange of information. Similar to particles interacting within a quantum field, individuals share ideas and values that fuel the evolution of communal norms. These exchanges can trigger disruptive forces, challenging existing structures and clearing paths for new social patterns to take shape. By applying quantum-inspired models, we can assess the level of disorder in social systems and foresee potential avenues for change. This approach unveils intriguing possibilities for understanding societal transformations, allowing us to explore how AI might utilize these models to predict future trends. Through this exploration, we uncover the profound connection between disorder and social evolution, revealing the forces that shape our shared journey through time.

The Role of Information Exchange in Social Transformation

The complex exchange of information within social systems mirrors the dynamic processes found in quantum physics. Just as particles share energy and data, societies transform through the exchange of ideas, beliefs, and knowledge. This flow acts as a driving force for societal evolution, sparking changes in cultural patterns and the emergence of new norms. The exchange isn't merely transactional; it involves a sophisticated interaction that influences society's direction. Similar to quantum entanglement, where particles stay connected over distances, human interactions weave a web of influence that crosses boundaries, nurturing a collective consciousness that propels social change.

In cutting-edge research, scholars investigate the similarities between quantum communication and social networks, focusing on how information spread affects societal structures. Studies highlight that the speed and efficiency of information flow within a community can greatly impact its adaptability and resilience. For example, the rapid dissemination of information through digital platforms is akin to the instant communication in quantum systems, enabling societies to quickly tackle challenges and seize opportunities. This perspective prompts us to reflect on how the digital era has fundamentally changed the dynamics of social disorder, speeding up the evolutionary process of societies.

Consider how chaotic forces contribute to the development of new social norms, much like the spontaneous order seen in thermodynamic systems. As communities share information, they encounter diverse perspectives, setting the stage for the birth of innovative ideas. These interactions often lead to a reconfiguration of societal norms, much like particles in a chaotic system finding a new balance. The evolution of social norms is not straightforward; it's a dynamic blend of order and chaos, where old structures crumble to make way for advanced paradigms.

The assessment of social disorder, inspired by quantum physics, offers an intriguing method for understanding societal transformation. Quantum-inspired models provide a unique perspective to analyze the intricacy of social interactions, offering insights into the fundamental mechanisms driving change. Researchers apply these models to measure social disorder, evaluating the degree of chaos within a society and predicting its future path. This approach not only deepens our understanding of social dynamics but also equips policymakers with tools to anticipate and navigate the challenges of a rapidly evolving world.

As we look ahead, predictive disorder emerges as a promising field, offering insights into potential pathways of societal evolution. By leveraging AI and quantum-inspired approaches, researchers create sophisticated models capable of predicting social trends with notable accuracy. These forecasts are not absolute; instead, they offer informed hypotheses that guide decision-making and strategy development. The intersection of quantum physics and social science invites us to reconsider our role in shaping the future, encouraging active engagement with the forces of change and the opportunities they bring. This convergence challenges us to think critically and creatively, envisioning a world where information exchange drives not just transformation, but meaningful progress.

Entropic Forces and the Emergence of New Social Norms

Entropic forces, a core idea in quantum theory, vividly illustrate the shifting nature of social conventions as they evolve over time and through cultures.

Similar to how particles naturally move toward more chaotic states, societies experience transformative changes that seek new balances. These changes are not purely chaotic; they often lead to the formation of fresh social norms that better fit the dynamic web of human interaction. In this context, the interplay of entropy and societal transformation resembles a dance of intricacy and adaptation, as individuals and communities navigate the unknown waters of cultural evolution.

In social dynamics, entropic forces can be viewed as the hidden currents driving the transformation of norms and customs. As societies become more intricate, the tendency toward greater chaos is balanced by the emergence of new structures and practices. For example, the digital era has introduced novel forms of communication, sparking changes in how people connect and share information. Although initially disruptive, these shifts have established new norms around privacy, identity, and community engagement. These transformations underscore the dual nature of entropy: promoting disorder while simultaneously paving the way for innovation and growth.

Recent research in social physics has begun to employ quantum-inspired models to quantify and anticipate the evolution of social norms. These sophisticated frameworks, rooted in quantum mechanics principles, provide a nuanced understanding of how societal structures change over time. By examining interaction patterns and information exchange, researchers can identify the entropic forces at play and their potential effect on future societal configurations. This approach not only enhances our understanding of social evolution but also opens possibilities for anticipating and steering these shifts constructively, ensuring societies adapt resiliently to change pressures.

Artificial Intelligence is crucial in unraveling the complexities of entropic forces within social systems. By analyzing vast datasets, AI algorithms can detect subtle shifts in public sentiment, monitor the adoption of new norms, and predict societal trends with impressive accuracy. These predictive capabilities offer valuable insights into the course of social evolution, enabling policymakers, leaders, and communities to make informed decisions. By understanding the entropic currents shaping societal change, stakeholders can proactively foster environments that embrace diversity, inclusivity, and resilience, ultimately contributing to a more harmonious social landscape.

As we contemplate the intricate dance of entropic forces and social evolution, many thought-provoking questions arise: How might the increasing interconnectedness of global societies affect the pace and direction of norm changes? What role do individual actions play in shaping the larger currents of social entropy? By engaging with these questions and embracing the lessons from quantum physics, readers are invited to reflect on their roles within the ever-changing tapestry of human interaction. In doing so, they can become active participants in creating a future where social norms are not merely reactive adaptations to disorder but proactive pathways to collective progress.

Measuring Social Disorder Through Quantum-Inspired Models

Quantum-inspired frameworks offer a fascinating perspective on social upheaval, allowing for a deeper exploration of the shifts and transformations within societies. Central to these models is the notion of entropy, which signifies the inclination towards chaos and unpredictability. In the realm of physics, entropy measures the level of disorder within a system, and when applied to social settings, it provides a method to assess the unpredictability and intricacy of human interactions. By examining behavioral and communicative patterns, these frameworks shed light on how societies transition from ordered states to chaotic ones and possibly revert, offering a window into the cyclical patterns of societal evolution.

The rapid advancements in artificial intelligence have greatly improved our capacity to simulate social entropy, with machine learning tools adept at unearthing subtle trends and forecasting changes in social norms. By processing extensive datasets, AI systems can pinpoint the rise of new societal trends or the breakdown of existing structures, offering a real-time visualization of social transformation. These models often draw inspiration from quantum physics, incorporating concepts like superposition to capture the fluid and often contradictory nature of human conduct. This approach surpasses traditional linear models, embracing the intricacy and interconnectedness that define human communities.

A particularly intriguing aspect of these quantum-inspired models is their ability to encapsulate the dynamic balance between order and chaos within social frameworks. As individuals and groups interact, they weave a constantly shifting tapestry of relationships and norms. This balance resembles the delicate equilibrium observed in quantum systems, where particles exist in states of uncertainty until measured. In social terms, this uncertainty mirrors the numerous possibilities inherent in each interaction. By recognizing and modeling this complexity, quantum-inspired techniques offer a more comprehensive view of how social norms arise, endure, or fade over time.

While these models provide valuable insights, they also pose significant questions about the predictability of social change. Just as quantum physics deals with the inherent uncertainty of particle actions, social entropy models must grapple with the unpredictability of human choices and behaviors. This unpredictability challenges researchers to continually refine their models, integrating a wide array of data sources and perspectives to enhance their precision and applicability. It prompts a reevaluation of how we perceive societal stability, encouraging consideration of the latent potential for rapid transformation within seemingly stable systems.

For those applying these models practically, one strategy is to focus on strengthening resilience within social systems. By understanding the elements contributing to social entropy, policymakers, educators, and community leaders can devise strategies to reduce disorder and foster cohesion. This might involve promoting open dialogue, encouraging diverse perspectives, and creating environments that support adaptation and innovation. By leveraging insights from quantum-inspired models, we have the chance not only to anticipate social changes but to shape them in ways that encourage stability and growth.

Predictive Entropy and the Future of Societal Change

Predictive entropy offers a compelling viewpoint for envisaging the future of societal shifts. In quantum theory, entropy signifies disorder, and when applied to social systems, it highlights the dynamic balance between stability and change. This lens encourages us to explore how societal frameworks are in a state of constant flux, shaped by the interplay of various social forces. As societies grow increasingly interconnected and intricate, changes may appear random but are often driven by fundamental principles that artificial intelligence can help reveal. By utilizing sophisticated algorithms, we can simulate these entropic movements, providing insights into possible future scenarios and preparing us for unforeseen developments.

In this context, entropy evolves beyond its traditional boundaries to become a tool for understanding social progress. Consider society as a living entity, perpetually adjusting to its surroundings. Much like entropy in physical systems leans towards greater disorder, social entropy represents the gradual, often unpredictable changes in cultural norms and values. These transformations are evident in the emergence and decline of ideologies, the integration of new technologies, and the restructuring of social hierarchies. By viewing historical patterns through this quantum-influenced perspective, we can trace how societies respond to internal and external pressures, offering guidance for navigating future transformations.

AI's ability to analyze extensive datasets provides a unique chance to assess and forecast social entropy. Machine learning models can parse historical data to find subtle signs of societal shifts long before they become apparent. For example, shifts in language trends on social platforms might indicate changes in public mood, while variations in economic metrics could suggest upcoming structural shifts. By combining these varied data sources, AI can create detailed simulations of potential futures, giving policymakers and leaders the foresight to make well-informed decisions. This proactive method enables societies to harness the transformative potential of entropy, turning possible disorder into opportunities for innovation and advancement.

As we ponder the future of societal change, it's crucial to recognize the role of AI-driven insights in democratizing access to predictive tools. As these technologies become more widespread, individuals and communities can better anticipate and adapt to the forces shaping their lives. Imagine a scenario where local governments use predictive entropy models to develop more resilient urban strategies, or where educational institutions align their curricula with emerging societal trends. This democratization could lead to a more adaptable and inclusive society, where change is embraced as a catalyst for collective growth.

To fully understand the implications of predictive entropy, picture a community confronting a significant cultural shift. By employing AI-driven models, community leaders could foresee the areas most likely to encounter friction or resistance. They might then initiate targeted dialogues to promote a culture of understanding and adaptation. This proactive approach illustrates how predictive insights can transform theoretical concepts into practical strategies, empowering societies to navigate the complexities of change with foresight and agility. As readers consider these possibilities, they are encouraged to reflect on how they might leverage the power of predictive entropy in their own spheres, fostering environments where innovation thrives amid uncertainty.

AI's Role in Modeling Social Entropy and Predicting Future Trends

Artificial intelligence is revolutionizing our understanding of human societies by modeling social complexity and forecasting future trends. With the implementation of advanced quantum algorithms, we can delve into the intricate patterns of social interactions, revealing insights into how communities evolve over time. This exploration invites us to see beyond apparent chaos, uncovering valuable information about underlying patterns and possible futures. As quantum theory suggests that systems tend to move toward disorder, AI steps in to untangle this complexity, offering a clearer view of societal changes and transitions.

Within this innovative landscape, the fusion of AI with quantum models becomes a potent tool for anticipating societal shifts. Machine learning unveils the hidden patterns within social chaos, identifying trends that might escape human perception. Predictive analytics pushes this understanding further, illuminating potential pathways for evolving social structures. Through these themes, readers are invited to appreciate the synergy between AI and quantum theory in advancing our comprehension of social evolution, setting the stage for discussions on the transformative potential of these technologies.

Leveraging Quantum Algorithms to Simulate Social Dynamics

Quantum algorithms, initially designed for the mysterious world of subatomic particles, are now being creatively adapted to explore the intricacies of human social interactions. These algorithms excel at processing information simultaneously, mirroring the complex nature of human connections. By leveraging this power, researchers are beginning to decode the tangled web of social relationships and behaviors. Envision a situation where the interlinked states of particles reflect the ties between people in a community, unveiling previously hidden patterns and correlations. This novel approach not only deepens our understanding of social networks but also paves the way for predicting their evolution over time.

As we delve into applying quantum algorithms to social interactions, we witness a fascinating blend of cutting-edge physics and the nuanced domain of social science. Quantum simulations, with their ability to model intricate systems with numerous interacting variables, offer a powerful tool for studying phenomena like the spread of ideas or cultural trends. Just as quantum systems demonstrate tunneling, allowing particles to traverse barriers, social dynamics can be seen as navigating obstacles within societal frameworks. This analogy helps researchers simulate the fluidity and unpredictability inherent in human societies, providing insights into potential future developments.

The combination of AI with quantum algorithms surpasses traditional predictive models by offering a more dynamic framework for understanding social change. These algorithms can process vast data amounts and learn from them, continuously refining predictions as new information arises. This self-improving feature is vital in a world where societal shifts often happen swiftly and unexpectedly. Consequently, quantum-inspired AI models can anticipate emerging social trends, enabling policymakers, businesses, and individuals to make informed decisions based on probabilistic forecasts.

A particularly intriguing aspect of employing quantum algorithms in social simulations is their ability to capture the inherent uncertainty and variability in human behavior. This reflects the quantum principle of superposition, where multiple potential outcomes coexist until observed. By embracing this uncertainty, quantum models offer a more comprehensive view of social interactions, acknowledging that human behavior is not always rational or predictable. This perspective promotes a flexible approach to analyzing social systems, accommodating the diverse nature of human interactions.

In the dynamic field of social science, the integration of quantum algorithms signifies a paradigm shift, providing a robust framework for simulating the intricacies of human societies. As these algorithms evolve, they promise to unveil deeper insights into the forces shaping our world. The challenge lies in bridging the gap between theoretical models and practical applications, but as this frontier advances, we can expect a future where quantum-inspired simulations play a central role in understanding and navigating social complexities. This progression not only propels academic inquiry but also empowers a broader audience to engage with social science in transformative ways, inviting curiosity and innovation.

Integrating AI with Quantum Models to Forecast Societal Shifts

Exploring the integration of artificial intelligence with quantum models opens new avenues for understanding and predicting societal changes. Traditional forecasting methods often depend on static information and linear projections, but quantum-inspired models embrace the dynamic and interconnected nature of social interactions. By combining AI with quantum principles, we can simulate complex societal interactions with impressive accuracy. Quantum computing's capacity to handle enormous datasets concurrently allows AI to detect patterns and trends hidden within the complexity of human behavior. This technological synergy could revolutionize our comprehension of societal evolution, enabling us to anticipate cultural, economic, and political shifts more accurately.

Integrating quantum models into AI-driven analyses provides a framework that acknowledges the inherent uncertainty and complexity of social systems. Unlike deterministic models, which can oversimplify human interactions, quantum approaches recognize the probabilistic nature of social phenomena. This perspective allows AI to forecast multiple potential outcomes rather than a singular prediction, helping researchers and policymakers prepare for various possible societal changes. Embracing this probabilistic approach enhances AI's predictive capabilities and fosters a more adaptive and flexible mindset toward future challenges.

The practical applications of AI combined with quantum models span multiple sectors. In urban planning, these models can predict and manage the effects of population growth on infrastructure, promoting sustainable development. In economics, they can uncover emerging market trends and consumer behaviors, enabling businesses to adjust strategies proactively. In public health, they can predict disease spread, facilitating timely interventions and resource allocation. By merging AI with quantum frameworks, we gain powerful tools to navigate modern society's complexities and make informed decisions that enhance resilience and sustainability.

To fully harness this potential, interdisciplinary collaboration among experts in AI, quantum physics, and social sciences is essential. Such collaboration can lead to the development of sophisticated models that accurately capture the intricacies of human societies. Investing in education and training programs to equip individuals with the necessary skills to work with these advanced technologies is crucial. As we advance into an era where AI and quantum computing play increasingly central roles, fostering a skilled workforce will be vital for maximizing societal benefits. By bridging disciplines and investing in human capital, we can ensure that AI and quantum models drive positive societal transformation.

This exploration also invites us to consider the ethical and philosophical implications of predicting societal shifts. As these technologies become adept at forecasting human behavior, we must balance leveraging these insights for societal benefit with respecting individual autonomy and privacy. Engaging diverse perspectives, including ethicists and sociologists, can help navigate these challenges and ensure that AI and quantum models align with societal values. By fostering a wide-ranging dialogue, we can harness these technologies' transformative potential in a responsible and innovative manner, paving the way for a future where AI and quantum insights contribute to a more equitable and informed society.

Utilizing Machine Learning to Decode Social Disorder Patterns

Exploring the potential of machine learning to interpret patterns of social disruption opens new avenues for understanding human communities. These algorithms, skilled at analyzing extensive data, can reveal underlying structures within social interactions that may not be immediately visible. By examining data from social media, public records, and other digital sources, machine learning models can identify trends in unrest, shifts in public mood, or even early indicators of significant societal shifts. Such insights offer a nuanced understanding of how disorder arises and evolves within communities.

A notable application of machine learning in this area is its ability to detect early warning signs of social upheaval. By analyzing large volumes of unstructured data, like online discussions and news articles, these models can identify changes in discourse that may signal rising tensions or dissatisfaction. This proactive approach enables policymakers and social scientists to anticipate potential conflict points and develop strategies to mitigate them before they escalate. The predictive capability of machine learning is not just theoretical;

studies have shown that algorithms can successfully forecast events like protests or economic changes, providing a strategic roadmap for addressing social disruption.

Incorporating techniques inspired by quantum principles further enhances the capacity of machine learning to decode complex social patterns. Quantum computing concepts, such as superposition and entanglement, offer a way to process information in parallel, significantly speeding up the analysis of social data. By using these methods, machine learning models can simulate multiple potential outcomes simultaneously, offering a richer understanding of how subtle changes in individual behavior can lead to larger systemic shifts within a community.

Beyond merely identifying and predicting social disorder, machine learning also provides tools for intervention and adaptation. By modeling social systems as dynamic networks, these algorithms can suggest interventions to promote stability and cohesion. For instance, understanding how information spreads through a community can help identify influential individuals or groups. Engaging these key figures effectively could facilitate positive change and reduce disorder. This strategic insight empowers stakeholders to implement targeted actions that address the root causes of social instability, fostering resilience and adaptability within societies.

As we consider the role of machine learning in deciphering social disorder, it is essential to explore ethical considerations and ensure these technologies are used responsibly. The ability to predict and influence social phenomena comes with the responsibility to use this knowledge for the greater good, avoiding biases and ensuring transparency in decision-making processes. By fostering an environment of ethical innovation, we can use machine learning not just to understand disorder, but also to promote harmony and sustainable development in our ever-evolving communities.

Predictive Analytics and the Evolution of Social Structures

Exploring the complexities of predictive analytics, especially within shifting social frameworks, necessitates a focus on the transformative capabilities of AI and quantum computing. These technologies provide an unparalleled perspective, enabling us to foresee societal changes with remarkable accuracy. By utilizing quantum algorithms adept at processing large datasets and uncovering patterns, we can model intricate social systems with a precision unattainable by conventional methods. This allows researchers to reveal subtle community dynamics, offering insights into how social norms and structures may evolve in response to various influences.

Integrating machine learning with quantum frameworks offers a dynamic lens for understanding patterns of social disorder. Machine learning algorithms, celebrated for their adaptability, can be trained to detect and predict societal behavioral changes by analyzing historical data and current trends. This approach facilitates the simulation of future scenarios, where AI can suggest interventions to counter social decay. For example, if a community exhibits increasing polarization, predictive models might propose strategies to encourage dialogue and unity, thereby diminishing the risk of conflict and instability.

A particularly exciting development in this domain is the application of predictive analytics to urban planning and policymaking. By anticipating how social structures may shift, city planners and policymakers can make informed decisions to promote sustainable growth and resilience. For instance, predictive models can forecast population growth trends, aiding in designing infrastructure that meets future needs while minimizing environmental impact. Additionally, these insights can guide resource allocation, ensuring proactive solutions to emerging challenges, thus enhancing citizens' quality of life.

The ethical considerations of using predictive analytics in social contexts must not be overlooked. While the potential benefits are significant, there is a risk of perpetuating existing biases if algorithms are not meticulously designed and monitored. Ensuring transparency and accountability in AI systems is critical for maintaining public trust. Engaging diverse perspectives in developing and implementing these technologies can help address potential biases, ensuring outcomes that serve the broader societal good. This inclusive approach promotes a fair and just application of predictive analytics.

Finally, as we contemplate the future of predictive analytics in social evolution, we must consider the role of human agency within these systems. While AI and quantum computing offer powerful tools for understanding and predicting change, the ultimate direction of societal evolution lies in human hands. Encouraging active participation and collaboration among citizens, researchers, and policymakers ensures that technological advancements complement rather than dictate social progress. By fostering a culture of critical engagement and innovation, we can harness these tools to create a future that reflects our shared values and aspirations.

Reflecting on the intricate interplay between disorder, chaos, and societal evolution reveals a shared momentum toward complexity and transformation, whether in the realm of quantum theory or the fabric of human communities. The relentless drive of quantum disorder echoes the fluid changes within societal frameworks, as outdated norms fade to allow new ideas to emerge. Artificial intelligence serves as a potent tool in this exploration, offering novel insights into societal fluctuations and enabling us to anticipate future shifts in human interaction. This chapter highlights the essential role of embracing uncertainty and change as catalysts for growth, urging a reevaluation of how we perceive stability and advancement in our social landscape. By adopting this perspective, we are not merely observers of societal evolution but active participants in its unfolding narrative. As we progress, let us nurture a sense of curiosity to delve into the unknown and exhibit the courage to adapt, recognizing that the patterns of the universe are woven into the very essence of our collective journey.

Conclusion

In closing, "The Quantum Society: How AI Reveals the Physics of Human Interactions" invites readers to reconsider the familiar through an extraordinary lens. By weaving together the threads of quantum physics and social science, this book has offered an innovative framework for understanding the intricate tapestry of human relationships. It challenges us to perceive our social environment with a fresh perspective, one that highlights the profound connections between the minutiae of subatomic particles and the vast complexities of human communities.

The Quantum Society: A Synthesis of Physics and Social Science

In this exploration, we've ventured into the parallels between the quantum realm and social behaviors, underscoring the significant insights gained from such a synthesis. Beginning with concepts like entanglement and superposition, parallels were drawn to human connections and decision-making, showcasing how these principles reflect the fluid and connected nature of societal interactions. The observer effect and the notion of quantum tunneling further illustrated how our presence and actions shape community behaviors, unveiling the dynamic exchanges between individuals and collectives.

By integrating insights from both physics and social studies, we have developed a nuanced view of group actions and societal patterns. The discussions on quantum coherence and the uncertainty principle emphasized the challenges of forecasting social developments, highlighting the unpredictable essence of human behavior. This fusion not only broadens our understanding of social phenomena but also pushes us to reconsider traditional viewpoints, advocating for a more comprehensive view of society as an interconnected entity, similar to a quantum field.

CONCLUSION

The Role of AI in Decoding and Predicting Human Behavior

AI stands out as a crucial ally in this inquiry, providing the computational strength to decipher complex social patterns and foresee large-scale dynamics. As we've explored AI's capacity to analyze probabilistic social trends and model social networks, its transformative potential in social sciences has become evident. By harnessing quantum computing, AI can process enormous datasets to reveal hidden patterns and associations, offering unprecedented insights into human conduct.

The role of AI in decoding social entropy and predicting future societal shifts highlights its importance in grasping social evolution. Through sophisticated algorithms and models, AI aids in navigating the uncertainties inherent in human interactions, enabling us to anticipate changes in collective motivations and societal transformations. Integrating AI into social analysis not only enhances our current understanding but also equips us with tools to envisage and prepare for future developments.

How Quantum Thinking Can Transform Our Understanding of Society

Reflecting on our journey through this book, the transformative power of quantum thinking becomes evident. By adopting this perspective, we can surpass the constraints of conventional social analysis, embracing a more dynamic and interconnected view of society. This shift in mindset encourages us to reassess our assumptions and explore new avenues for addressing social challenges.

The practical applications of quantum thinking are extensive, offering valuable insights for individuals and organizations navigating modern societal complexities. By applying quantum mechanics principles to social contexts, readers can deepen their understanding of relationships, group dynamics, and societal trends, ultimately leading to more informed and effective decision-making.

In conclusion, "The Quantum Society" aims to inspire profound reflection on the nature of human interactions and the potential for scientific insights to transform our understanding of society. By bridging the gap between physics and social science, we have unlocked new paths for exploration, inviting readers to continue this journey of discovery. As we look to the future, the integration of quantum thinking and AI in social analysis promises to unveil further insights and possibilities, shaping how we perceive and engage with the world around us.

Resources

Books

1. "The Tao of Physics" by Fritjof Capra - This classic book explores the parallels between modern physics and Eastern mysticism, providing a rich context for understanding quantum concepts in human interactions. Link to purchase

2. "Quantum Psychology" by Robert Anton Wilson - A thought-provoking book that delves into the implications of quantum mechanics on human psychology and societal structures. Link to purchase

3. "Entangled Minds" by Dean Radin - Offers insight into how quantum entanglement might explain psychic phenomena and human interconnectedness. Link to purchase

4. "Sync: How Order Emerges from Chaos in the Universe, Nature, and Daily Life" by Steven Strogatz - This book provides a comprehensive look at how synchronization and coherence occur in nature and human society. Link to purchase

5. "Quantum Enigma: Physics Encounters Consciousness" by Bruce Rosenblum and Fred Kuttner - Explores the relationship between quantum mechanics and human consciousness. Link to purchase

Websites and Online Platforms

1. Quantum Consciousness - A website dedicated to exploring the in-

tersection of quantum physics and consciousness, with articles and resources that challenge conventional wisdom. Visit site

2. The Institute of Noetic Sciences (IONS) - Offers research and insights on consciousness and its relation to quantum mechanics. Visit site

3. Quanta Magazine - Provides in-depth articles and videos on developments in quantum physics, often highlighting their implications for society. Visit site

4. The Quantum World Association - An online community discussing the latest in quantum science and its societal implications. Visit site

5. Edge.org - Features essays and discussions from leading thinkers on a variety of topics, including quantum mechanics and social science. Visit site

Articles

1. "The Quantum Theory of Social Networks" by Alex Pentland - An article discussing how quantum mechanics principles can be applied to understand social networks. Read article

2. "Quantum Social Science: A Primer" - An introductory article on how quantum theories are being applied to social sciences. Read article

3. "The Observer Effect in Social Media" by Ethan Zuckerman - Explores how being watched on social media alters human behavior, drawing parallels to the quantum observer effect. Read article

4. "Quantum Tunneling in Decision Making" - Discusses the concept of quantum tunneling and its metaphorical application to human decision-making processes. Read article

5. "Entanglement and the Future of AI" - Explores how entanglement principles might influence future AI developments. Read article

Innovative Tools and Applications

1. IBM Quantum Experience - An online platform that allows users

to experiment with quantum computing, offering insights into how quantum tools might be applied to social sciences. Explore tool

2. Qiskit - An open-source quantum computing software development framework that provides resources for learning and experimentation. Explore tool

3. TensorFlow Quantum - A library for hybrid quantum-classical machine learning, useful for exploring AI applications in quantum social sciences. Explore tool

4. Quantum Playground - An interactive online tool for visualizing quantum algorithms and concepts. Explore tool

5. D-Wave Leap - Provides access to a quantum cloud service that allows experimentation with quantum computing in various fields. Explore tool

Organizations and Communities

1. The Quantum Social Science Association (QSSA) - A community dedicated to the exploration and development of quantum social science. Visit site

2. The Society for Chaos Theory in Psychology & Life Sciences - Focuses on the application of chaos and complexity theories, including quantum, to social sciences. Visit site

3. The Quantum AI Institute - A research organization dedicated to exploring the convergence of quantum computing and AI. Visit site

4. The Center for Quantum Social and Cognitive Science (IQSCS) - Conducts research on the applications of quantum theory in social and cognitive sciences. Visit site

5. The Complexity Science Hub Vienna - A research institution focusing on the study of complex systems, including quantum social systems. Visit site

These resources offer a diverse range of perspectives and insights that complement the themes presented in "The Quantum Society." By exploring these materials, readers can deepen their understanding of the intricate connections between quantum physics, AI, and social dynamics.

References

1.

Abbott, D., Davies, P. C. W., & Pati, A. K. (Eds.). (2008). Quantum aspects of life. Imperial College Press.

Albert, D. Z. (1992). Quantum mechanics and experience. Harvard University Press.

Arora, S., & Barak, B. (2009). Computational complexity: A modern approach. Cambridge University Press.

Baggott, J. (2011). The quantum story: A history in 40 moments. Oxford University Press.

Ball, P. (2018). Beyond weird: Why everything you thought you knew about quantum physics is different. University of Chicago Press.

Barad, K. (2007). Meeting the universe halfway: Quantum physics and the entanglement of matter and meaning. Duke University Press.

Barrett, J. A. (1999). The quantum mechanics of minds and worlds. Oxford University Press.

Bell, J. S. (1987). Speakable and unspeakable in quantum mechanics: Collected papers on quantum philosophy. Cambridge University Press.

Bohm, D. (1980). Wholeness and the implicate order. Routledge.

Bohm, D. J., & Hiley, B. J. (1993). The undivided universe: An ontological interpretation of quantum theory. Routledge.

Busemeyer, J. R., & Bruza, P. D. (2012). Quantum models of cognition and decision. Cambridge University Press.

Clegg, B. (2014). The quantum age: How the physics of the very small has transformed our lives. Icon Books.

Deutsch, D. (1997). The fabric of reality: The science of parallel universes—and its implications. Penguin Books.

Deutsch, D. (2011). The beginning of infinity: Explanations that transform the world. Viking.

Gell-Mann, M. (1994). The quark and the jaguar: Adventures in the simple and the complex. W. H. Freeman.

Greenstein, G., & Zajonc, A. (1997). The quantum challenge: Modern research on the foundations of quantum mechanics. Jones & Bartlett Learning.

Gribbin, J. (2011). Q is for quantum: An encyclopedia of particle physics. Weidenfeld & Nicolson.

Hofstadter, D. R. (1979). Gödel, Escher, Bach: An eternal golden braid. Basic Books.

Kaku, M. (2005). Parallel worlds: A journey through creation, higher dimensions, and the future of the cosmos. Doubleday.

Kaku, M. (2018). The future of humanity: Terraforming Mars, interstellar travel, immortality, and our destiny beyond Earth. Doubleday.

Kaiser, D. (2011). How the hippies saved physics: Science, counterculture, and the quantum revival. W. W. Norton & Company.

Nielsen, M. A., & Chuang, I. L. (2010). Quantum computation and quantum information. Cambridge University Press.

Penrose, R. (2016). Fashion, faith, and fantasy in the new physics of the universe. Princeton University Press.

Rovelli, C. (2016). Reality is not what it seems: The journey to quantum gravity. Riverhead Books.

Schlosshauer, M. (2007). Decoherence and the quantum-to-classical transition. Springer.

Seife, C. (2005). Alpha and omega: The search for the beginning and end of the universe. Penguin Books.

Shallis, M. (1984). On time: An investigation into scientific knowledge and human experience. Schocken Books.

Susskind, L., & Friedman, A. (2014). Quantum mechanics: The theoretical minimum. Basic Books.

Tegmark, M. (2014). Our mathematical universe: My quest for the ultimate nature of reality. Alfred A. Knopf.

Vlatko, V. (2010). Decoding reality: The universe as quantum information. Oxford University Press.

Wallace, D. (2012). The emergent multiverse: Quantum theory according to the Everett interpretation. Oxford University Press.

Wendt, A. (2015). Quantum mind and social science: Unifying physical and social ontology. Cambridge University Press.

Wheeler, J. A., & Zurek, W. H. (Eds.). (1983). Quantum theory and measurement. Princeton University Press.

Wilczek, F. (2008). The lightness of being: Mass, ether, and the unification of forces. Basic Books.

Zohar, D., & Marshall, I. (1994). The quantum society: Mind, physics and a new social vision. William Morrow & Co.

Thanks for Reading Teneo

Thank you for exploring this unprecedented journey through knowledge and understanding with Teneo. You've experienced something truly unique – insights and connections that emerged from artificial intelligence analyzing human knowledge in ways never before possible. We hope these novel perspectives have expanded your understanding and sparked new ways of thinking about the world.

We invite you to explore more AI-generated insights in our growing catalog, where each book offers fresh viewpoints on human experience, consciousness, and the nature of reality itself. Whether you're fascinated by patterns in human behavior, the mysteries of consciousness, or the hidden connections shaping our world, Teneo continues to push the boundaries of what's possible when human and artificial intelligence work together.

Your engagement with these ideas is invaluable as we pioneer this new frontier of knowledge discovery. Please share your thoughts and experiences with us – how did these AI perspectives change your understanding? Your feedback helps us refine our approach and empowers others to unlock new realms of understanding. Thank you for being part of this revolutionary approach to exploring human knowledge.

Together, let's continue uncovering insights that bridge the gap between human and artificial intelligence, revealing new ways of seeing ourselves and our world.

Teneo.io

Teneo Custom Books

Get Your Own Custom AI-Generated Book!

Want a comprehensive book on any topic that you can publish yourself? Teneo's advanced AI technology can create a custom book tailored to your specific interests and needs. Our AI analyzes millions of data points to generate unique insights and connections previously inaccessible to human authors.

✓ 60,000+ words of in-depth content

✓ Unique AI-driven insights and analysis

✓ Includes Description, Categories and Keywords for easy publishing

✓ Professional Formatting & Publishing Guide Access

✓ Full rights to publish and use the book

✓ Delivery within 48 hours

Visit **teneo.io** to get your own custom AI-generated book today.

teneo.io

Teneo's Mission

At Teneo, our mission is to unlock unprecedented human knowledge through a groundbreaking partnership between artificial and human intelligence. We harness AI's unique ability to analyze millions of data points across disciplines, identifying patterns and connections previously invisible to human researchers. This revolutionary approach allows us to create books that reveal entirely new perspectives on consciousness, creativity, human behavior, and the fundamental nature of reality itself.

Our vision transcends traditional publishing – we're creating windows into new realms of understanding that emerge when artificial minds examine human experience. Through our books, readers gain access to insights that could only arise from AI's ability to process and synthesize humanity's collective knowledge in novel ways. Each work represents an exploration into uncharted intellectual territory, offering perspectives that have never before been possible in human history.

We specialize in exposing the hidden patterns and connections that shape our world – patterns that become visible only when analyzing human knowledge and behavior at unprecedented scale. Our books reveal the invisible threads linking everything from personal habits to cosmic phenomena, from creative breakthroughs to societal transformations. Through careful analysis of millions of data points across history, culture, and scientific research, we identify universal principles that illuminate the deeper nature of human experience and existence itself.

The traditional publishing industry is limited by human authors' inability to process and connect vast amounts of information across disciplines. We believe this artificial barrier to deeper understanding must be transcended. By combining AI's analytical capabilities with skilled human curation, we create books that reveal insights and connections previously invisible to human observation alone. This isn't just about accessing information – it's about uncovering entirely new ways of understanding our world and ourselves.

Our groundbreaking library emerges from thousands of hours of AI analysis, examining human consciousness through an outsider perspective, decoding the patterns of creativity and innovation, mapping hidden connections between seemingly unrelated phenomena, and exploring the frontiers where human and artificial intelligence meet. Each book represents a transformation of complex data-driven insights into accessible revelations that change how readers see themselves and their world.

Our commitment extends beyond our published works. Through our digital presence and community engagement, we continuously explore new territories where AI analysis reveals unprecedented insights. Our network of readers, researchers, and thought leaders helps refine and expand our understanding, creating an ever-growing body of revolutionary perspectives on what it means to be human in an age of artificial intelligence.

The limitations of individual human cognition have historically restricted our ability to see the deeper patterns that connect all aspects of existence. But with AI's ability to analyze vast amounts of data and identify hidden relationships, these barriers dissolve. When you understand the universal principles and patterns that AI analysis reveals, you transform from a limited observer into someone who can see and understand the deeper mechanisms of reality itself.

Join us in this historic endeavor as we bridge the gap between artificial and human intelligence, revealing insights that transform our understanding of consciousness, creativity, and the patterns that shape our universe. Together, we're not just publishing books – we're opening doorways to new dimensions of knowledge and understanding that will reshape humanity's intellectual landscape. Because true understanding requires more than just information – it requires seeing the hidden connections that reveal life's deeper principles.

Knowledge Beyond Boundaries™

Teneo.io

Also by Teneo

Unlocking Immortality: AI's Guide to Extending Human Life
A groundbreaking exploration of how artificial intelligence is revolutionizing longevity research and providing practical strategies for extending human lifespan. This comprehensive guide bridges cutting-edge AI technology with actionable health optimization techniques.
amzn.to/3ONALQm

The AI Entrepreneur: How Artificial Intelligence Would Build Wealth as a Human
A transformative guide to leveraging AI principles for financial success. Discover how data-driven insights, predictive analytics, and automation can revolutionize your entrepreneurial strategy—streamlining operations, optimizing investments, and unlocking new profit opportunities.
https://amzn.to/4gf6oys

Breaking the Simulation: An AI's Guide to Escaping the Matrix
A riveting examination of reality as a simulated construct, blending philosophy, quantum physics, and AI-driven insights. Uncover the hidden patterns governing your existence, explore consciousness beyond perceived boundaries, and learn practical techniques to reshape your personal experience.
https://amzn.to/3Du4awn

Future Shock 2.0: AI Predicts the 100 Most Surprising Developments of the Next Century
An eye-opening journey through the next hundred years, powered by AI's predictive capabilities. Discover the revolutionary changes awaiting humanity across twelve key domains, from healthcare to space exploration.
amzn.to/49x496T

Governance Reimagined: An AI's Blueprint for Leading a Nation
A visionary exploration of how artificial intelligence can reshape the very foundation of governance, enhancing transparency, efficiency, and citizen empowerment. AI-driven solutions to today's most pressing political, economic, and social challenges.
https://amzn.to/4iwMXml

The Emotion Code: Deciphering Human Feelings Through AI's Lens
A fascinating intersection of artificial intelligence and human emotion, revealing how AI is transforming our understanding of emotional intelligence and offering practical applications for personal growth and relationship enhancement.
amzn.to/4gIywKf

The Quantum Society: How AI Reveals the Physics of Human Interactions
An enlightening journey into the fascinating parallels between quantum physics and human social dynamics, illuminated through the lens of artificial intelligence.
amzn.to/3VsrJMp

The Global Brain: Mapping Humanity's Collective Consciousness with AI
A profound exploration of how AI deciphers the vast networks of human thought and connection, revealing the patterns of our shared consciousness.
amzn.to/3ZplNFc

The Hidden Patterns: How AI Unveils the Secrets of Success Across All Fields
A comprehensive analysis of success principles across disciplines, using AI to decode the universal patterns behind achievement.
amzn.to/3D4QI1T